APPLICATION GUIDE FOR ABSORPTION COOLING/REFRIGERATION USING RECOVERED HEAT

This publication was prepared under ASHRAE Research Project 773-RP
by Dorgan Associates, Inc.
Madison, Wisconsin

APPLICATION GUIDE FOR ABSORPTION COOLING/ REFRIGERATION USING RECOVERED HEAT

Chad B. Dorgan
Steven P. Leight
Charles E. Dorgan

American Society of Heating, Refrigerating and Air-Conditioning Engineers, Inc.

ASHRAE STAFF

Frank M. Coda
Executive Director/
Publisher

W. Stephen Comstock
Director
Communications/Publications

Mildred Geshwiler
Editor
Special Publications

Michelle Moran
Associate Editor
Special Publications

Estelle Bowen
Administrative
Assistant/Secretary
Special Publications

Stefanie Frick
Secretary
Special Publications

Scott Zeh
Manager
Publishing Services

ISBN 1-883413-26-5

Copyright 1995
American Society of Heating, Refrigerating and Air-Conditioning Engineers, Inc.
1791 Tullie Circle, NE / Atlanta, GA 30329

Printed in the USA

ASHRAE has compiled this publication with care, but ASHRAE has not investigated, and ASHRAE expressly disclaims any duty to investigate, any product, service, process, procedure, design, or the like that may be described herein. The appearance of any technical data or editorial material in this publication does not constitute endorsement, warranty, or guaranty by ASHRAE of any product, service, process, procedure, design, or the like. ASHRAE does not warrant that the information in the publication is free of errors, and ASHRAE does not necessarily agree with any statement or opinion in this publication. The entire risk of the use of any information in this publication is assumed by the user.

Acknowledgments

This Guide was developed through a cooperative effort of Dorgan Assoiates, Inc., and absorption machine experts who reviewed various drafts and provided both suggestions and input data. The input of the ASHRAE Project Monitoring Subcommittee, chaired by Dr. Horacio Perez-Blanco (Penn State University) and members Steve Petty (Columbia Gas Service Corporation), Reinhard Radermacher (University of Maryland), Uwe Rockenfeller (Rocky Research Corporation), and Bill Ryan (Gas Research Institute) was instrumental in achieving a successful Guide.

Over 20 people reviewed the draft versions. The following deserve special thanks for their detailed comments, suggestions, and supplemental information: Mike Byars (The Trane Company), Dan Erdmann (Affiliated Engineers, Inc.), Jim Furlong (York International Corporation), Paul Guevara (The Trane Company), Asama Ibrahim (University of Rhode Island), Jay Kohler (York International Corporation), Frank Lamphere (Henry Vogt Machine Company), Werner Malewski (Deutsche Babcock-Borsig), Skip McCullough (Baltimore Aircoil Company), Kevin McGahey (American Gas Cooling Center), Mike Pawelski (The Trane Company), Bill Plzak (The Trane Company), Jim Porter (The Trane Company), Doug Rector (United Technology Carrier), Bob Reimann (United Technology Carrier), Jim Rozanski (Cain Industries), Jim Shepherd (Lewis Energy Systems), Walter Smith, Jr. (Energy Technology Services International), and George Vicatos (University of Cape Town).

Dorgan Associates' staff members who deserve recognition for their contributions include Marion McGavock, technical writer, whose input and assistance were invaluable in accomplishing changes to graphics and for maintaining uniformity throughout the Guide. Anthony Eggert, Ryan Schmid, Mike Armstrong, and Steve Parsons assisted with research and development of illustrations and calculations. Thanks to Joan Dorgan for cheerfully assisting in material preparation and typing. A special thanks to Marisue Quigley for an outstanding job of work processing, proofreading, and final editing and formatting.

About the Authors

Chad B. Dorgan, P.E., is a senior engineer and **Charles E. Dorgan, Ph.D., P.E.**, is president of Dorgan Associates, Inc., Madison, Wisconsin. At the time this project was conducted, **Steven P. Leight** was with Dorgan Associates.

Contents

CHAPTER 1 — INTRODUCTION

1.1 Background 1

1.2 Benefits of Absorption Systems 2

1.3 Absorption System Applications 3

    1.3.1 Heat Recovered from a Cogeneration System 3

    1.3.2 Heat Recovered from an Industrial Process 3

1.4 Basic Terminology 5

    1.4.1 Effects 5

    1.4.2 Stages 5

    1.4.3 Coefficient of Performance 5

1.5 Types of Indirect-Fired Absorption Systems 7

1.6 Reference Sources 8

    1.6.1 American Society of Heating, Refrigerating and Air-Conditioning Engineers, Inc. 8

    1.6.2 Gas Research Institute 9

    1.6.3 American Gas Cooling Center 9

    1.6.4 American Gas Association 9

    1.6.5 Absorption Machine Vendors 9

		1.6.6 Gas Utility Companies 9

        1.6.7   Trade Magazines 10

        1.6.8   University Research Theses and Papers 10

        1.6.9   Others 10

        1.6.10 Essential References for Absorption Applications 11

CHAPTER 2 — REQUIRED SYSTEM CHARACTERISTICS

    2.1   Heat Source 13

            2.1.1   Temperature of the Source Heat Stream 14

                  2.1.1.1   High Grade 15

                  2.1.1.2   Medium Grade 16

                  2.1.1.3   Low Grade 16

            2.1.2   Flow Rate of Recovered Heat Stream 16

            2.1.3   Chemical Composition of the Source Heat Stream 17

            2.1.4   Intermittency of Recovered Heat Stream Temperature and Flow 17

    2.2   Cooling Load 17

    2.3   Heat Sink 18

    2.4   Matching Heat Sources, Cooling Loads, and Heat Sinks 19

    2.5   Cooling Estimates 22

    2.6   Summary 22

CHAPTER 3 — ECONOMICS OF ABSORPTION SYSTEMS

3.1 Costs 25

3.2 First Costs 26

    3.2.1 Absorption Machine 26

    3.2.2 Heat Rejection Equipment 26

    3.2.3 Heat Recovery Equipment 28

    3.2.4 Auxiliary Equipment 28

3.3 Operating Costs 28

    3.3.1 Absorption Machine Energy Use 28

    3.3.2 Energy Penalty of Heat Recovery Equipment 30

    3.3.3 Cooling Water Pump and Cooling Tower Fan Energy Use 30

3.4 Maintenance Costs 33

3.5 Summary of Costs 33

CHAPTER 4 — HEAT RECOVERY

4.1 Methods of Heat Recovery 37

    4.1.1 Steam 37

        4.1.1.1 Steam Generation for Single-Effect Absorption Machines 38

        4.1.1.2 Steam Generation for Double-Effect and Ammonia-Absorption Machines 38

    4.1.2 Hot Water 38

4.2 Heat Recovery Equipment 38

	4.2.1	Shell-and-Tube Heat Exchangers (Liquid or Steam to Hot Water) 39
	4.2.2	Exhaust Gas Boilers (exhaust gas to steam or hot water) 40
	4.2.3	Steam Separators 41
4.3	Cogeneration 42
	4.3.1	Gas Turbines 42
	4.3.2	Steam Turbines 43
	4.3.3	Reciprocating Engines 43

CHAPTER 5 — THE ABSORPTION PROCESS

5.1	The Vapor-Compression Cycle 47
5.2	The Absorption Cycle 49
5.3	Absorbent-Refrigerant Solutions 51
	5.3.1 Types of Absorption Systems 52
	5.3.2 Solution Concentration Levels 52
	5.3.3 Other Solutions 53
5.4	Chapter Lisiting of Different Systems 53
	5.4.1 Lithium Bromide-Water System 53
	5.4.2 Aqueous Ammonia System 53

CHAPTER 6 — LITHIUM BROMIDE-WATER ABSORPTION CYCLE

| 6.1 | PTX Equilibrium Chart .... 55 |

	6.1.1	Solution Temperature	56
	6.1.2	Solution Concentration	56
	6.1.3	Refrigerant Temperature	57
	6.1.4	Refrigerant Vapor Pressure	57
6.2	Single-Effect System	57
	6.2.1	Single-Effect Absorption Process	57
	6.2.2	Single-Effect Absorption Machine Configuration	60
6.3	Double-Effect System	61
	6.3.1	Double-Effect Absorption Process	62
	6.3.2	Double-Effect Machine Configuration	66
6.4	Comparison of Single- and Double-Effect Systems	66
6.5	Individual Components	69
	6.5.1	Pumps	69
	6.5.2	Absorber Section	69
	6.5.3	Solution Heat Exchanger	70
	6.5.4	Expansion Device	71
	6.5.5	Evaporator	72
	6.5.6	Purge System	72
	6.5.7	Cooling Water System	73
	6.5.8	Controls	73
		6.5.8.1 Crystallization	74

		6.5.8.2 Capacity Control 75
6.6	Part-Load Operation 76	
6.7	Chemical Additives 80	
	6.7.1 Corrosion Inhibitors 80	
	6.7.2 Mass/Heat Transfer Enhancement 81	
	6.7.3 pH Balancers 81	
	6.7.4 Testing of Solution 81	
6.8	Maintenance 82	

CHAPTER 7 — AQUEOUS AMMONIA ABSORPTION CYCLE

7.1	Graphical Representation of AAR Cycle 83
7.2	One-Stage System 84
7.3	Multistage Systems 86
7.4	Individual Mechanical Components 88
	7.4.1 Pumps 88
	7.4.2 Absorber Section 88
	7.4.3 Evaporator Section 89
	7.4.4 Solution Heat Exchanger 89
	7.4.5 Generator/Rectifier 90
	7.4.6 Ammonia Condenser 91
	7.4.7 Expansion Device 91

	7.4.8	Cooling Water System 92
	7.4.9	Controls 92
7.5	Chemical Additives 92	

CHAPTER 8 — MATCHING HEAT SOURCES, COOLING LOAD CAPABILITY AND HEAT REJECTION WITH AN ABSORPTION MACHINE

8.1	Introduction 95	
8.2	Determine Cooling Requirements and Heat Source 96	
8.3	Select Type of Absorption Machine to Use 96	
8.4	Calculate Recoverable Heat 97	
	8.4.1	Gas Streams 97
	8.4.2	Liquid Streams 99
	8.4.3	Engine Jackets 99
	8.4.4	Backpressure Steam Turbine 100
	8.4.5	Mass Flow of Recovered Heat 100
8.5	Determine the Cooling Output Attainable 101	
8.6	Size the Absorption Machine 102	
8.7	Calculate Heat Rejection Requirements 102	
8.8	Calculate Electric Power Requirements 103	
8.9	Resize Machine with Other Options 103	
8.10	Determine Part-Load and Nonconcurrent Load Operation 103	

	8.10.1 Excess Recovered Heat	103
	8.10.2 Insufficient Recovered Heat	104
8.11	Automation of Application Procedure	104

CHAPTER 9 — LiBr ABSORPTION MACHINE APPLICATIONS

9.1	Introduction	107
9.2	Example Description	107
9.3	Determining Cooling Requirements and Heat Source	108
9.4	Select Type of Absorption Machine to Use	109
9.5	Calculate Recoverable Heat	110
9.6	Determine Amount of Cooling Output	112
9.7	Size the Absorption Machine	114
	9.7.1 Steam Pressure Correction	114
	9.7.2 Cooling Water and Chilled-Water Temperature Correction	116
9.8	Heat Rejection Requirements	118
9.9	Electrical Power Requirements	119
	9.9.1 Absorption Machine Electrical Energy Use	119
	9.9.2 Cooling Water Pump Electrical Energy Use	121
	9.9.3 Cooling Tower Fan Electrical Energy Use	122
9.10	Consider Other Options	123
9.11	Nonconcurrent Loads	126

9.11.1 Mismatched Loads 126

9.11.2 Part-Load Conditions 126

9.12 Cost Estimates 129

CHAPTER 10 — AAR MACHINE APPLICATIONS

10.1 Introduction 131

10.2 Example Description 131

10.3 Calculate Recoverable Heat 132

10.4 Determine Amount of Cooling Output 136

10.5 Other Sizing Parameters 137

CHAPTER 11 — OTHER TECHNOLOGIES

11.1 Triple Effect 139

11.2 GAX Cycle 140

11.3 Heating 140

11.3.1 Heat Amplifier 140

11.3.2 Temperature Amplifier 140

11.4 Combination Heating and Cooling 142

APPENDIX A — PROPERTY CHARTS, TABLES, AND CONVERSIONS 143

APPENDIX B — ALGORITHMS FOR CHARTS AND COMPUTER PROGRAM 151

APPENDIX C — INPUT INSTRUCTIONS FOR COMPUTER PROGRAM 157

APPENDIX D — EXISTING APPLICATIONS LISTING 161

GLOSSARY 177

BIBLIOGRAPHY 181

CHAPTER 1

Introduction

1.1 BACKGROUND

ASHRAE commissioned this application guide in response to the need for a comprehensive reference manual for the application of indirect-fired absorption machines. *Indirect fired* is defined as an absorption machine that uses heat recovered from another process or heat cycle machine. Direct-fired absorption machines are not covered in the guide. This document will help engineers and owners to

- become familiar with the requirements of an absorption system,
- evaluate indirect-fired absorption machines for specific applications,
- evaluate the potential of available recovered heat sources, and
- select the most economical system.

The purpose of this guide is to provide the reader with information on the successful application of absorption machines. The actual design of an absorption system is covered in detail in other publications, which are listed in the bibliography of this guide. However, common areas of concern in the design of absorption machines are briefly discussed in the guide.

The application of absorption for the purpose of cooling has been a viable technology for more than a century. This guide contains the information required to properly incorporate absorption technology into a wide range of applications, including commercial space cooling, central plant cooling, and industrial applications where recoverable heat is available.

Absorption systems can provide an economical use for recovered heat energy. This guide assumes that a heat source is available, economically appropriate, and cannot be eliminated through modification of the process or power cycle. There is no higher economic use of the potential recovered heat than whenever there is a cooling or refrigeration requirement that can be satisfied with the recovered heat.

The guide is structured to provide a logical progression from vapor-compression technology to absorption technology. It is assumed that the reader of this guide understands the basic principles of refrigeration and the fundamentals of vapor-compression systems. The guide introduces the concepts of absorption cooling using the terminology of vapor-compression systems. This information is expanded to enable the reader to apply absorption technology to many situations with available recovered heat.

The guide is organized into 11 chapters. Chapter 1, *Introduction*, provides basic background on the application of absorption systems and terminology. Chapter 2, *Required System Characteristics*, presents information on the fundamental requirements that make indirect-fired absorption systems feasible. Chapter 3, *Economics of Absorption Systems*, provides a discussion on the economics of absorption applications. Chapter 4, *Heat Recovery*, discusses the types and sources of heat and how to recover and use the heat in absorption applications. Chapter 5, *The Absorption Process*, presents general information on the absorption process. Chapter 6, *Lithium Bromide-Water Absorption Cycle*, provides a detailed discussion of lithium bromide-water (LiBr) absorption machines. Chapter 7, *Aqueous Ammonia Absorption Cycle*, provides a detailed discussion of water-ammonia absorption refrigeration machines (AAR). Chapter 8, *Matching Heat Sources, Cooling Load, and Heat Rejection Capability with an Absorption Machine*, provides detailed procedures for applying indirect-fired absorption machines to specific applications. Chapter 9, *LiBr Absorption Machine Applications*, presents specific information for the application of LiBr machines. Similarly, Chapter 10, *AAR Machine Applications*, presents specific information for the application of aqueous ammonia absorption machines. Chapter 11, *Other Technologies*, discusses technologies that incorporate absorption cycles for the heating, ventilating, and air-conditioning (HVAC) industry.

As this is an applications guide and not a design guide, detailed sizing procedures are not included.

References and bibliographies, where appropriate, are provided at the end of each section, and a comprehensive bibliography is included at the end of the guide. Definitions of terms used in the guide are included in the glossary.

1.2   BENEFITS OF ABSORPTION SYSTEMS

The first absorption machine was developed in the mid-nineteenth century by a Frenchman, Ferdinand Carre. Since this first machine, the popularity of absorption systems has risen and fallen due to economic conditions and technology breakthroughs from competing technologies. The benefits of absorption systems have remained constant and include the following:

- absorption systems have lower electrical needs compared to vapor-compression systems,
- absorption units are quiet and vibration free,
- recovered heat can be utilized to power the absorption refrigeration cycle,
- absorption units pose no threat to global environmental ozone depletion and may have less impact on global warming than most other options, and
- absorption units are economically attractive when fuel costs are substantially less than electric costs (typically, if fuel cost is 12% to 20% of electrical cost) (Smith 1994; Furlong 1994).

1.3 ABSORPTION SYSTEM APPLICATIONS

A refrigeration system uses an energy input to produce a cooling effect. In vapor-compression systems, the energy input is mechanical work, usually an electric motor. In absorption systems, the energy input is heat.

If a facility or process has the potential to recover heat, an absorption system can use the recovered heat to generate cooling. Typically, heat for absorption cooling is recovered from a process and used in process cooling or to reduce the temperature of the surrounding environment by use of an HVAC system. Figure 1-1 shows a variety of applications for absorption machines.

1.3.1 Heat Recovered from a Cogeneration System

Applying indirect-fired absorption machines with cogeneration systems is currently the most common approach. Heat recovered from the exhaust gas stream and the engine coolant system is used to power the absorption machine. When the heat from the cogeneration plant is used to power an absorption machine, this not only meets the cooling load but also reduces the peak electric demand on the system.

The cooling load on the absorption machine will vary depending on the local requirements. Several common applications include

- cooling occupied spaces (HVAC system),
- cooling inlet air for gas turbines (which increases gas turbine capacity and efficiency),
- providing refrigeration to a food processing plant or for ice production, and
- cooling an industrial process.

1.3.2 Heat Recovered from an Industrial Process

The application of absorption systems has typically been economical for many industrial facilities with both large heat rejection and cooling requirements. The heat source and cooling load in an industrial facility vary, depending on the processes. Examples of possible heat sources include

- exhaust gas from a drying process, such as in a paper plant;
- hot water produced in a wastewater treatment facility;
- gas vapor produced in an oil refinery from a distillation process;
- low-pressure steam from a backpressure steam turbine; and
- incinerator exhaust/coolant water.

APPLICATION GUIDE FOR ABSORPTION COOLING/REFRIGERATION USING RECOVERED HEAT

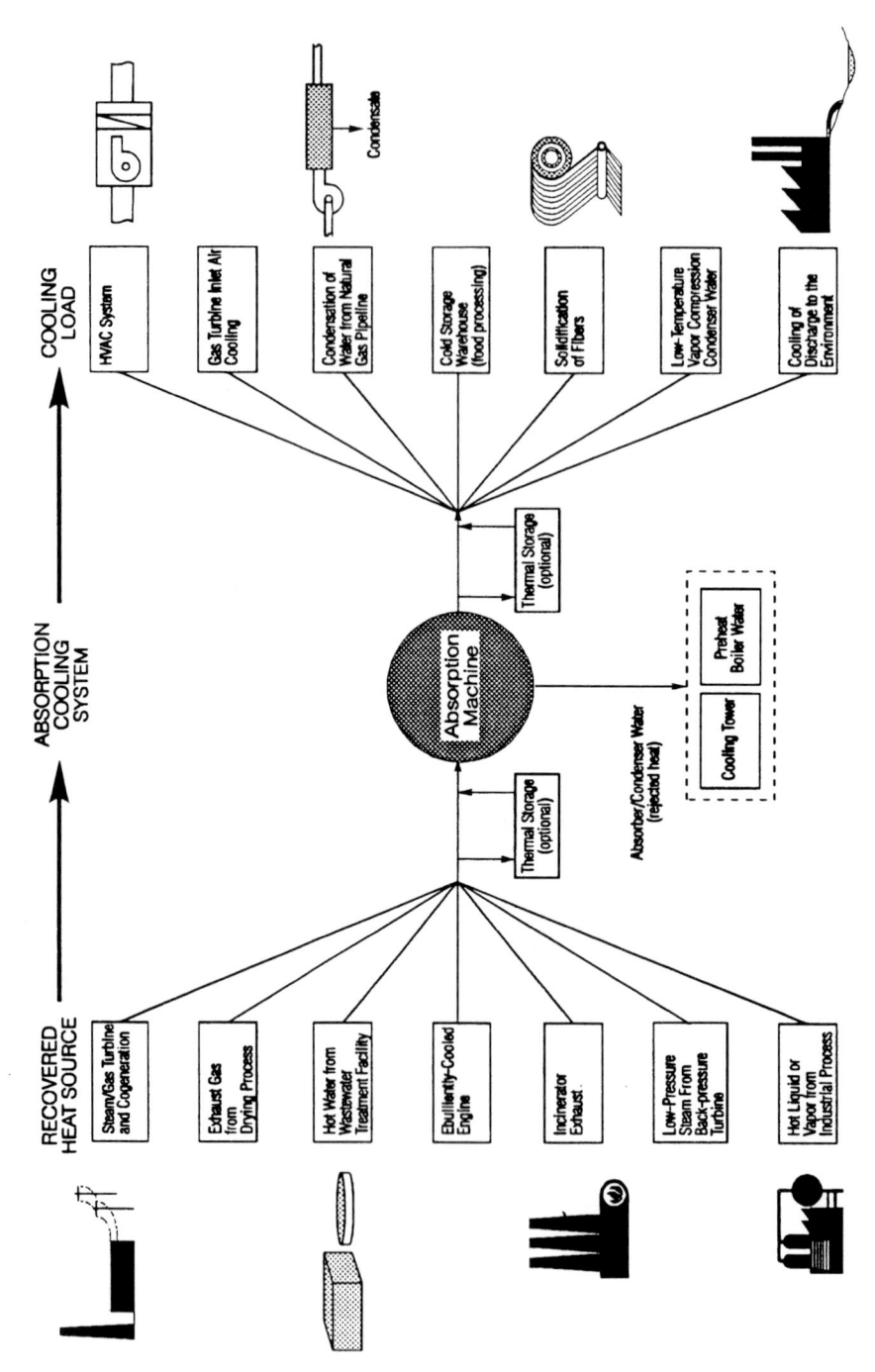

Figure 1-1. Absorption machine applications.

The potential cooling loads in an industrial facility vary as much as the heat sources. Examples include cooling for a dewaxing process, solidifying synthetic fibers, cooling of condenser water for a low-temperature vapor-compression machine, spot cooling for personal comfort (HVAC), and cooling a wastewater stream prior to introduction into the environment.

1.4 BASIC TERMINOLOGY

This section presents terminology to clarify operating and efficiency concepts relating to absorption machines. Knowledge of basic terminology helps reduce confusion, especially when comparing alternative systems. Key terms used in this guide are defined in the glossary.

1.4.1 Effects

Absorption machines are categorized either by the number of *effects* or by the number of *stages*. In this guide, an effect refers to the number of times the input heat is used by the absorption machine, either directly or indirectly. In a single-effect system, the input heat is used only once. In a double-effect system, the input heat is used twice, and in a triple-effect system the input heat is used three times. A detailed explanation of effects is found in chapter 5, *The Absorption Process*.

1.4.2 Stages

In this guide, a stage refers to the number of evaporator/absorber pairs at different temperatures in an absorption machine. A single-stage system has a single evaporator/absorber operating at a constant or primary temperature. A two-stage system has two evaporator/absorber pairs, each operating at a different temperature.

An application of a two-stage system is a refrigerated warehouse with a cooler requiring –20°F (–29°C) and a freezer requiring –40°F (–40°C) (Holldorff and Malewski 1987).

1.4.3 Coefficient of Performance

The coefficient of performance[1] (COP) of an absorption machine is a ratio used to rate the energy transfer effectiveness of the machine. For refrigeration, this is typically interchanged with efficiency. COP, defined in Equation 1-1, is the amount of cooling delivered by a machine divided by the amount of energy input required to produce the cooling (ASHRAE 1993). This is a rating ratio of energy transferred to energy input. The COP does not include auxiliary electricity to operate the pumps and fans.

---

[1] The COP referenced in this Guide is for cooling applications only.

$$COP = \frac{Cooling\ Delivered}{Energy\ Input\ Required} \tag{1}$$

The COP for a single-effect system is typically two-thirds that of a double-effect system. For a single-effect system, COPs range from 0.6 to 0.7. For a double-effect system, COPs range from 0.9 to 1.2. A double-effect machine makes more efficient use of the input energy. For low-temperature absorption systems, COPs range from 0.1 to 0.8 (ASHRAE 1988; Bogart 1981; Vicatos 1994).

For vapor-compression systems, the COPs range from 2 to 6.75. While the COPs for vapor-compression systems are higher than those for absorption systems, vapor-compression systems typically require a higher grade of energy to power them (electricity versus gas or recovered heat). Therefore, when comparing the COP of an absorption cycle to that of a vapor-compression cycle, it is important to account for the differences in the cost of the energy required by each. In a vapor-compression cycle, the COP is determined by the amount of mechanical work required to compress the vapor. In an absorption cycle, the COP is determined by the amount of heat required to operate the cycle.

Energy required to operate condenser fans, cooling towers, circulating pumps, and other auxiliary energy use is not included in refrigeration COP ratings. However, these energy costs must be included in the full economic analysis of the system. Also, the controls required to achieve acceptable part-load and nonpeak design efficiencies must be accounted for in the life-cycle cost analysis.

Since the cost to produce the mechanical work needed to provide one ton (3.5 kW_T) of cooling in a vapor-compression cycle is typically greater than the cost to recover the heat required by an absorption system to produce one ton (3.5 kW_T) of cooling, absorption systems can often compete economically with vapor-compression systems. Variations in local utility rates and specific site information will affect the economic analysis.

As an example, compare a vapor-compression chiller with a COP of 5.5 to a double-effect absorption unit with a COP of 1.0. If the cost of electricity is more than 5.5 times greater than the value of the recovered heat, on a dollars per Btu ($/$kW_T$) basis, the absorption chiller will have an operating cost advantage (Furlong 1994). It is important to note that this is a simplified COP-based example and that the costs to operate auxiliaries such as cooling water pumps, cooling towers, and refrigerant pumps have not been included. By including the auxiliary energy use, the cost for electricity would have to increase from 5.5 to approximately 8.3 times greater for the absorption chiller to have an operating cost advantage. If the recovered heat has no value, then the COP has no impact for determining a simple economic decision.

Another factor to be considered when comparing COPs is the conditions under which the COPs were obtained. This guide uses COPs rated at standard conditions as published by the Air-Conditioning and Refrigeration Institute (ARI). The ARI guidelines for testing chillers (absorption or vapor compression) with cooling towers have the following conditions (ARI 1992a, 1992b, 1992c):

- supply cooling water temperature from the cooling tower: 85°F (29°C)
- chilled-water supply temperature: 44°F (7°C)
- chilled-water flow rate: 2.4 gpm/ton (0.043 L/s·kW$_T$)
- cooling water flow rate, single-effect: 3.6 gpm/ton (0.065 L/s·kW$_T$)
- cooling water flow rate, double-effect: 4.0 gpm/ton (0.072 L/s·kW$_T$)
- cooling water flow rate, vapor compression: 3 gpm/ton (0.054 L/s·kW$_T$).

If a comparison of different systems is to be accomplished, the cooling water temperature rise for the different types of systems at ARI conditions must be determined. From the above data, the cooling water temperature rise for single- and double-effect absorption machines and vapor-compression machines at ARI conditions is

- single-effect: 17°F (9°C),
- double-effect: 12°F (7°C), and
- vapor compression: 10°F (6°C).

1.5 TYPES OF INDIRECT-FIRED ABSORPTION SYSTEMS

Currently, there are two primary types of absorption systems commercially available. They are lithium bromide (LiBr)-water systems and aqueous (water)-ammonia refrigeration systems (AAR).

In a LiBr-water absorption system, the LiBr is the absorbent and water is the refrigerant. Typically, LiBr-water absorption systems are referred to as single- or double-effect machines. In an AAR system, the water is the absorbent and ammonia is the refrigerant.

The heat input requirements and the cooling-output-attainable temperatures vary between and within each system type. Figure 1-2 shows the general input and output parameters for indirect-fired absorption systems.

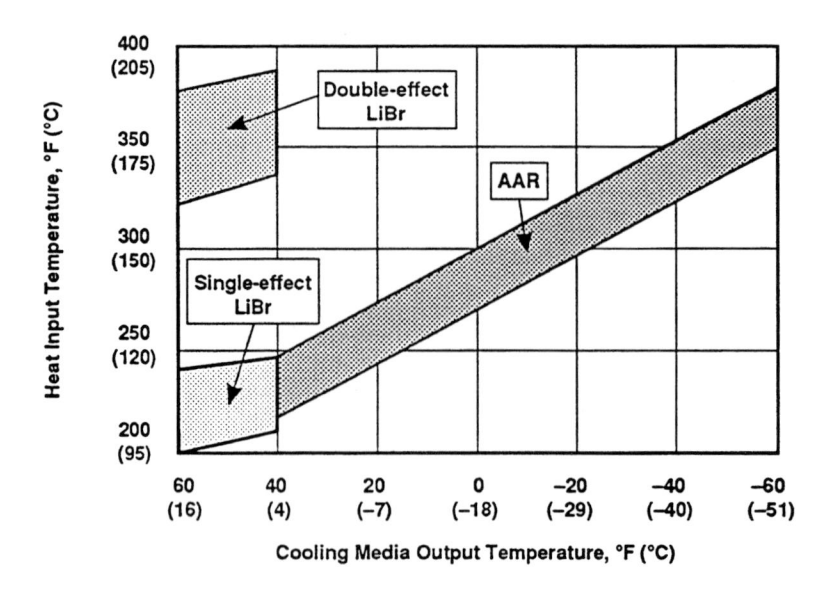

Figure 1-2. Absorption machine input/output.

1.6 REFERENCE SOURCES

Additional information on absorption systems is available from several sources. Among these are the American Society of Heating, Refrigerating and Air-Conditioning Engineers, Inc.; the Gas Research Institute; the American Gas Cooling Center; the American Gas Association; absorption machine vendors; gas utility companies; trade magazines; university research theses and papers; and others.

1.6.1 American Society of Heating, Refrigerating and Air-Conditioning Engineers, Inc.

The American Society of Heating, Refrigerating and Air-Conditioning Engineers, Inc. (ASHRAE) publishes technical papers and other publications and sponsors research covering a wide range of subject matters of interest to the HVAC&R industry. Technical papers presented at Society meetings are published biannually in *ASHRAE Transactions*. Further information on ASHRAE publications can be obtained by contacting ASHRAE Publications, 1791 Tullie Circle NE, Atlanta, GA 30329; phone (404) 636-8400; fax (404) 321-5478.

1.6.2 Gas Research Institute

The Gas Research Institute (GRI) provides funding for research of interest to the gas utility industry. GRI has recently funded research on absorption applications and has published a number of research reports. Reports are available through the United States Department of Commerce, National Technical Information Service, Springfield, VA 22161; phone (703) 487-4630; fax (703) 321-8547, or from the GRI Library, (312) 399-8354.

1.6.3 American Gas Cooling Center

The American Gas Cooling Center (AGCC) is a national trade association composed of North American gas utilities and manufacturers of gas cooling equipment. The primary goal of the AGCC is to act as a clearinghouse for information on gas equipment and the development of new industries. The AGCC can be contacted at 1515 Wilson Boulevard, Arlington, VA 22209; phone (703) 841-8542; fax (703) 841-8689.

1.6.4 American Gas Association

The American Gas Association (AGA) is a national trade association composed of North American natural gas utilities and pipeline companies. These firms, which deliver gas energy from the wellhead and various supplemental sources to the burner tip, account for 90% of all gas delivered by the regulated natural gas industry in the United States. The AGA acts as a clearinghouse on gas energy information, as a catalyst in technical and energy policy matters, and as a spokesperson for the regulated natural gas industry. Information can be obtained by contacting the AGA at 1515 Wilson Boulevard, Arlington, VA 22209-2470; phone (703) 841-8400; fax (703) 841-8687.

1.6.5 Absorption Machine Vendors

Manufacturers and installers of absorption equipment can provide valuable information concerning the use of their products in specific applications. While this information may be biased, discussions with advocates of several technologies will generally yield a useful comparison. Manufacturers' representatives may also offer assistance in sizing absorption machine and auxiliary equipment requirements. Annual resource issues of some of the trade magazines listed in section 1.6.7 provide information on manufacturers.

1.6.6 Gas Utility Companies

Local gas utility companies are often good sources of information on absorption systems. Many utilities have staff members devoted to encouraging application of absorption in their service territories. Although they typically concentrate on direct-fired systems, many utilities will help with indirect-fired systems.

1.6.7 Trade Magazines

Trade magazines often carry articles covering absorption installations and technical design issues related to absorption systems. The bibliography at the end of the guide contains citations for numerous articles from trade magazines. Additional references may be found in the most recent issues of magazines such as *ASHRAE Journal*, *Heating/Piping/Air Conditioning*, *Consulting-Specifying Engineer*, *Engineered Systems*, and *Air Conditioning and Refrigeration News*.

1.6.8 University Research Theses and Papers

Universities are a primary source of information on specific details of absorption machines. Individual university libraries can be accessed or organizations can be contacted to assist in locating information on specific topics.

1.6.9 Others

Other organizations that publish information on absorption machines include the following:

ARI — Air-Conditioning and Refrigeration Institute, 1501 Wilson Boulevard, Suite 600, Arlington, VA 22209; phone: (703) 524-8800.

ASME — American Society of Mechanical Engineers. Proceedings of the International Heat Pump Conferences. Held every three years. United Engineering Center, 345 East 47th St., New York, NY 10017; phone: (800) 843-2765; fax: (212) 705-7722.

DOE — Oak Ridge National Laboratory, Publication and Technical Information, Department of Energy, Oak Ridge, TN 37830; phone: (615) 576-1301; fax: (615) 576-2865.

IIAR — International Institute of Ammonia Refrigeration, 111 East Wacker Drive, Suite 600, Chicago, IL; phone: (312) 664-6610; fax: (312) 321-6869.

IIR — International Institute of Refrigeration. Publication: *International Journal of Refrigeration*. Published by Butterworth-Heinemann Ltd., Linacre House, Jordan Hill, Oxford OX2 8DP, UK. For information in the United States, contact: Frank M. Coda, Executive Director of ASHRAE, U.S. National Committee for the International Institute of Refrigeration; phone (404) 636-8400; fax: (404) 321-5478.

EPRI — Electric Power Research Institute, P.O. Box 10412, Palo Alto, CA 94303; phone: (415) 855-2000; fax: (415) 855-2954.

1.6.10 Essential References for Absorption Applications

All those pursuing the application of absorption machines will find the sources listed in the bibliography to be valuable.

REFERENCES

ARI. 1992a. *Standard 550, Centrifugal and rotary screw water-chilling packages*. Arlington, VA: Air-Conditioning and Refrigeration Institute.

ARI. 1992b. *Standard 560, Absorption water chilling and water heating packages*. Arlington, VA: Air-Conditioning and Refrigeration Institute.

ARI. 1992c. Standard 590, *Positive-displacement compressor water-chilling packages*. Arlington, VA: Air-Conditioning and Refrigeration Institute.

ASHRAE. 1988. *1988 ASHRAE handbook—Equipment*. Chapter 13, Absorption Cooling, Heating, and Refrigerating Equipment, pp. 13.2, 13.6, and 13.10. Atlanta: American Society of Heating, Refrigerating and Air-Conditioning Engineers, Inc.

ASHRAE. 1993. *1993 ASHRAE handbook—Fundamentals*, p. 1.4. Atlanta: American Society of Heating, Refrigerating and Air-Conditioning Engineers, Inc.

ASHRAE. 1994. *1994 ASHRAE handbook—Refrigeration systems and applications*. Chapter 40, Absorption Cooling, Heating, and Refrigeration Equipment, pp. 40.1–40.12. Atlanta: American Society of Heating, Refrigerating and Air-Conditioning Engineers, Inc.

Bogart, M. 1981. *Ammonia absorption refrigeration in industrial processes*, chapter 18, pp. 366–383. Houston: Gulf Publishing Co.

Furlong, J. 1994. Personal communication (letter).

Holldorff, G. M., and W. F. Malewski. 1987. *Operational experience in cogeneration plants with refrigeration supply to low-temperature cold storage* 1: 287–295. New York: American Society of Mechanical Engineers, International Gas Turbine Institute.

Smith, W. P., Jr. 1994. Personal communication (letter).

BIBLIOGRAPHY

AGCC. 1992. *Natural gas cooling equipment guide*. Arlington, VA: American Gas Cooling Center.

ASHRAE. 1988. *1988 ASHRAE handbook—Equipment*, chapter 13. Atlanta: American Society of Heating, Refrigerating and Air-Conditioning Engineers, Inc.

ASHRAE. 1992. *1992 ASHRAE handbook—HVAC systems and equipment*, chapters 7 and 8. Atlanta: American Society of Heating, Refrigerating and Air-Conditioning Engineers, Inc.

ASHRAE. 1994. *1994 ASHRAE handbook—Refrigeration systems and applications*, chapter 40. Atlanta: American Society of Heating, Refrigerating and Air-Conditioning Engineers, Inc.

ASME, Advanced Energy Systems Division. 1994. *Proceedings of the international absorption heat pump conference*, AES-vol. 31, January, New Orleans, LA.

Boyen, J. L. 1980. *Thermal energy recovery*, 2d ed. New York: John Wiley & Sons.

Grimm, N. R., and R. C. Rosaler, eds. 1990. *Handbook of HVAC design*, chapter 41, "Absorption Chillers," by N.J. Cassimatis. New York: McGraw-Hill.

CHAPTER 2
Required System Characteristics

This chapter is intended to give a brief introduction to the elements required for absorption systems and to help in determining whether an absorption system will be feasible in a given situation. In addition, information presented in this chapter can be used to estimate the size of required equipment.

There are three primary elements in every absorption system. These are the heat source, the cooling load, and the heat sink. The following three sections introduce general requirements for each of the primary elements. These sections are then followed by a section containing specific information for individual system types.

2.1   HEAT SOURCE

While a mechanical device, often an electric motor, drives a vapor-compression chiller, a heat source drives an absorption machine. Hot water, process steam, hot air, or hot products of combustion (exhaust gases) from the burning of natural gas or other fuel are all used as heat sources. Recovering steam or hot water from cogeneration facilities or industrial processes is a popular method for obtaining the driving force of absorption machines. Discussion in this section will focus on the requirements of a heat source for use in absorption systems.

When locating potential heat sources, it is necessary to evaluate several characteristics. The following variables are used to determine if a heat source is adequate (Boyen 1980):

- temperature of the source heat stream,
- flow rate of the recovered heat stream,
- chemical composition of the source heat stream, and
- intermittency of recovered heat stream temperature and flow.

The source heat stream is usually an exhaust gas, steam, or liquid stream that comes directly from a process or application. The recovered heat stream is typically heated by the source heat stream utilizing a heat exchanger. A heat exchanger is used to protect the absorption machine from contaminants and corrosive elements contained in the source heat stream. It is sometimes possible to use the source heat stream directly in the absorption machine if the source heat stream is clean. The equipment manufacturer should be contacted to determine if a specific source heat stream can be used directly.

2.1.1 Temperature of the Source Heat Stream

The average temperature of the source heat stream determines the temperatures or steam pressures that can be made available by using a heat recovery unit. Because of the varied heat input requirements of different types of absorption machines, the source heat stream temperature determines what types of absorption machines can be used. The desired temperature ranges for the heat input to the various absorption machines are shown in Table 2-1.

TABLE 2-1
Heat Input Requirements of Absorption Machines

Absorption Machine	Steam or Process Gas Temperature, °F (°C)	Hot Water or Process Fluid Temperature, °F (°C)
LiBr Single-Effect	225–250 (110–120)	240–300 (115–150)
LiBr Double-Effect	350–365 (175–185)	310–400 (155–205)
AAR Refrigeration	210–385 (100–195)	210–385 (100–195)

Note: This table is based on the range of manufacturers' data available in 1994.

As the input temperature to an absorption machine decreases, the cooling output also decreases. A heat source can be categorized as being one of three grades: high, medium, or low. The three grades are discussed below in detail (Rohrer and Kreider 1979). Typical processes and the corresponding recoverable temperatures are shown in Figure 2-1.

REQUIRED SYSTEM CHARACTERISTICS

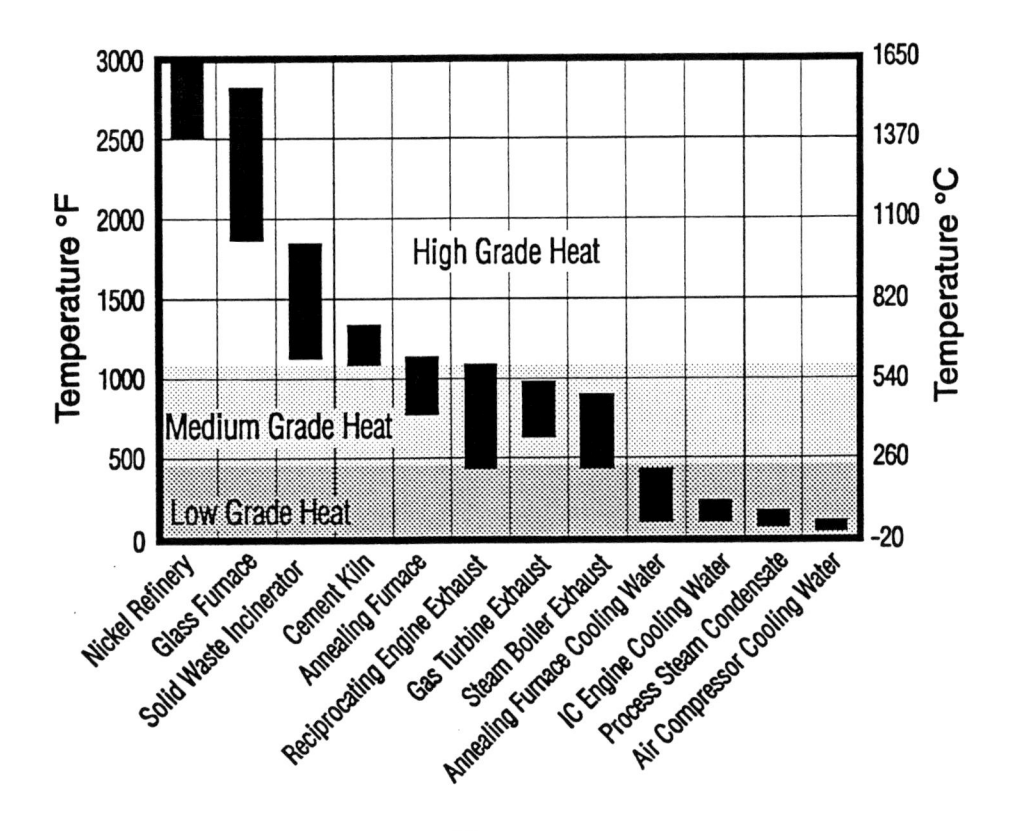

Figure 2-1 Process vs. Recovered Heat Temperature

2.1.1.1 High Grade

High-grade heat refers to heat at temperatures of more than 1,200°F (650°C). High-grade heat is usually a result of a direct-fuel-fired process. Typical processes that produce high-grade heat include glass manufacturing, metal refining, and solid and fume waste incineration.

High-grade heat can be applied directly to absorption machines. However, the most economical use of high-grade heat is to produce steam to typically drive a double-effect absorption machine or to drive an AAR absorption refrigeration machine. Steam pressures of more than 500 psig (3,550 kPa) can easily be generated by recovering high-grade heat.

15

2.1.1.2 Medium Grade

Medium-grade heat refers to heat between 450°F and 1,200°F (230°C and 650°C). Medium-grade heat is typically recovered from exhaust gases from a boiler, oven, or furnace and can be used to generate steam pressures of more than 200 psig (1,480 kPa).

For an absorption machine, the higher end medium-grade heat (higher than 800°F [425°C]) is used to drive double-effect machines or AAR absorption machines for low-temperature applications. Lower end medium-grade heat (less than 800°F [425°C]) is often used to produce low-pressure steam or hot water for use in a single-effect absorption machine.

2.1.1.3 Low Grade

Low-grade heat refers to heat of less than 450°F (230°C). Low-grade heat is the grade most commonly available in commercial and industrial processes. Sources of recovered low-grade heat are process steam condensate, process cooling water, steam from ebulliently cooled engines, and steam from backpressure steam turbines. Higher end low-grade heat (higher than 375°F [190°C]) is typically used to generate steam pressures from 150 to 200 psig (1,100 to 1,480 kPa).

Low-grade heat, in the form of steam or hot water, is sometimes used without a heat exchanger to drive single-effect absorption chillers. Recovered low-grade heat is also the heat source for absorption heat pumps. Absorption heat pumps are discussed briefly in chapter 11.

2.1.2 Flow Rate of Recovered Heat Stream

The flow rate and the temperature of a recovered heat stream determine if sufficient heat is available to produce the required cooling output. A typical single-effect absorption chiller will consume 18 pounds of steam per hour per ton of cooling (2.3 kg/h·kW_T). A typical double-effect absorption machine will consume 11 pounds of steam per hour per ton of cooling (1.4 kg/h·kW_T) (Smith et al. 1992).

A typical single-effect machine will consume varying amounts of hot water, depending on the initial temperature of the water and the temperature difference of the water across the absorption machine. The heat input for a single-effect machine using hot water ranges from 0.5 to 1.3 gpm/ton of cooling (0.009 to 0.02 L/s·kW_T). The greater the temperature difference, the lower the gpm/ton (L/s·kW_T) requirement. However, when lower temperature hot water is used, the cooling output decreases as compared to higher temperature hot water (Manufacturer 1975; Manufacturer 1989). Temperatures as low as 230°F (110°C) can be used economically.

A steady flow of recovered heat at a constant temperature is required by an absorption machine to maintain constant cooling output. Supplemental boilers may be added to provide additional heat. Radiators or load-balancing condensers can be used to dissipate excess heat. Installing a heat storage system can help match the heat source and cooling load profiles without the need for supplemental boilers or condensers.

2.1.3 Chemical Composition of the Source Heat Stream

The chemical composition of the source heat stream consists of the physical state of the stream (liquid or gas) and any contaminants in the heat stream.

The physical state of the source heat stream, whether gas or liquid, is important as it affects the type of heat recovery equipment used. A gas heat stream, such as the exhaust gas from an engine or a furnace, typically requires a gas-to-steam or a gas-to-water heat recovery unit, whereas a liquid heat stream, such as a liquid from an industrial process or the condensate from a steam turbine, typically requires a liquid-to-water or liquid-to-steam heat recovery unit.

Contaminants in the source heat stream can plug or foul heat exchangers, reducing the amount of heat recovered. Contaminants can also limit the type of heat exchange equipment that can be applied to a particular situation.

Methods are available to eliminate the problems associated with fouling agents present in a source heat stream. If the source heat stream is in a liquid state, sedimentation chambers or filters can be installed to remove contaminants. If the source heat stream is in a gas state, heat exchangers equipped with widely spaced bare tubes or steam- or air-operated soot blowers can be used.

Absorption machine manufacturers should be consulted for allowable fluids and contaminants if the source heat stream is to be used directly. Heat recovery manufacturers should be consulted to help in selecting the proper heat exchanger.

2.1.4 Intermittency of Recovered Heat Stream Temperature and Flow

The heat recovery equipment must be able to provide a consistent flow and temperature of recovered heat to the absorption machine in order to maintain constant cooling output. For projects with varying flow or temperature, installation of a supplemental heat source or heat-dissipating equipment to compensate for the difference between the recovered and required heat may be economical. Another option is to install heat thermal storage to provide heat for the absorption machine or cool thermal storage to match the cooling load.

2.2 COOLING LOAD

An absorption machine can be used for most of the same applications as a vapor-compression machine. Process cooling, space cooling and dehumidification, and low-temperature cooling and heating can all be accomplished using absorption machines that use recovered heat as the driving force. The primary topics for this guide are cooling and refrigerating applications. These are discussed below. Heating technologies are discussed in chapter 11.

Absorption machines are most commonly used for cooling applications. Lithium bromide-water absorption machines can provide chilled-water temperatures between 40°F and 100°F (4°C and 38°C) and are often used in air-conditioning applications. Aqueous ammonia absorption machines are typically used for refrigeration applications and can supply temperatures between 40°F and –60°F (4°C and –51°C).

When cooling requirements do not match the output of the absorption machine, cool thermal storage (water, ice, or eutectics) can be an economical option. Matching of the heat input, cool output, and the cooling load is critical in the many heat recovery systems where the heat must be dissipated at all times. Therefore, excess chilled water can be stored in tanks when space-cooling requirements are low and then used later used when cooling requirements are high. This type of arrangement is often used in cogeneration plants, where the heat source is relatively constant but the cooling requirements may be highly variable.

The COP of an absorption machine varies depending on the temperature of the recovered heat, the leaving temperature of the chilled fluid, and the type of absorption machine used. The COP for a single-effect LiBr machine ranges from 0.6 to 0.7 and ranges from 0.9 to 1.2 for a double-effect LiBr machine (ASHRAE 1992, 1993). For AAR systems, the COP ranges from 0.1 to 0.8 (Vicatos and Gryzagoridis 1994). The tradeoff between the higher COPs of a double-effect machine as compared to a single-effect machine is an increase in the cost of the machine and a higher temperature heat input requirement. The heat rejection for a double-effect absorption system is less due to the higher COPs. Absorption cooling, using no-cost recovered heat, can prove to be economical.

The COP has an impact on the first cost and size of an absorption machine. It is relatively meaningless for an economic evaluation with heat recovery equipment and is, therefore, not typically used for meaningful operating comparisons.

2.3 HEAT SINK

As with any type of cooling equipment, all absorption systems require some form of heat rejection. Typically, a cooling tower is used. However, other methods such as boiler feed water preheating can be used. Typically, preheating of boiler water, process water, or the use of the absorption machine cooling water for space comfort heating is limited to 110°F (43°C) to maintain reasonable refrigeration efficiency. As a rule of thumb, absorption chiller systems require between 1.5 to 2 times more system heat rejection than similar vapor-compression systems (Smith et al. 1992). This additional heat rejection is due to heat generated by the process of absorption.

If an absorption machine is replacing an existing vapor-compression system, then the existing cooling tower will not have enough capacity to reject all of the heat from the absorption system. The increased heat rejection requirements of an absorption machine also affect the amount of water treatment and the auxiliary loads (pumps and fans).

2.4 MATCHING HEAT SOURCES, COOLING LOADS, AND HEAT SINKS

The temperature and flow of the recovered heat affect the cooling output of an absorption machine. Also, the temperature difference of the cooling water across an absorption machine affect the cooling output of the machine. In this guide, cooling water is defined as the heat rejection fluid that removes the heat generated in the absorption cycle. Generally, the following guidelines and Figure 2-2 can be used to help understand the effect of changes in an absorption system input and output conditions.

- As the amount of heat input ($\dot{m}_3$, T_5 to T_6) is decreased, the cooling output (flow $\dot{m}_1$, T_1 to T_2) also decreases.
- With a constant entering temperature, as the cooling water temperature difference ($T_4 - T_5$) across an absorption machine is increased, the cooling output (flow $\dot{m}_1$, T_2 to T_1) decreases. The increase in the cooling water differential allows for a smaller cooling tower to be installed. This, however, may require the installation of a larger absorption machine (10% to 20%) to achieve the same cooling effect.
- If the cooling output (flow $\dot{m}_1$, T_2 to T_1) capacity is to remain constant, a higher temperature heat input (T_5) is required as the output temperature (T_1) is decreased.

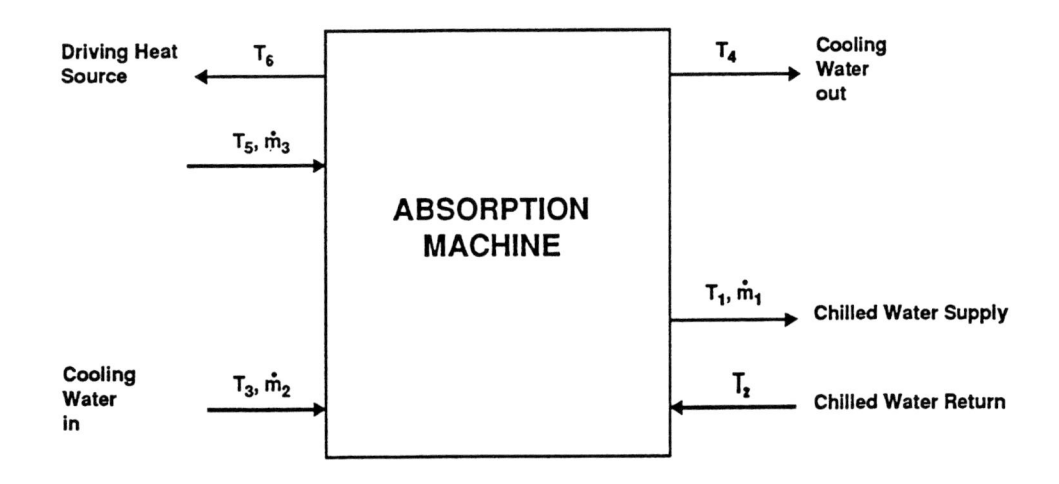

Figure 2-2 Absorption Machine Inputs and Outputs

Tables 2-2 and 2-3 have been developed using the application procedure introduced in chapters 8 through 10. Table 2-2 contains seven different system configurations using the same source heat stream and shows the impact of various conditions on absorption systems. Table 2-3 contains seven different system configurations with the same design cooling capacity.

TABLE 2-2
LiBr Absorption System Variations — Fixed Heat Source

Description	Units	Base Steam Case	Double-Effect Increased Steam Press.	Double-Effect Hot Water	Double-Effect Increased Cooling Water ΔT	Double-Effect Decreased CHW Temp.	Single-Effect Steam	Single-Effect Decreased CHW Temp.
Leaving CHW temperature	°F (°C)	42 (6)	42 (6)	42 (6)	42 (6)	40 (4)	42 (6)	40 (4)
Recovered heat quality	psig steam (kPa)	115 (894)	130 (998)	80 (653)	115 (894)	115 (894)	12 (184)	12 (184)
Amount recovered heat	Btuh (kW$_T$)	8,692,000 (2563)	8,572,000 (2528)	9,020,000 (2660)	8,692,000 (2563)	8,692,000 (2563)	10,065,000 (2967)	10,065,000 (2967)
Flow recovered heat	lb/hr (Kg/s)	8,094 (1.03)	8,042 (1.02)	8,219 (1.04)	8,094 (1.03)	8,094 (1.03)	10,734 (1.36)	10,734 (1.36)
Design cooling output	tons (kW$_T$)	800 (2810)	750 (2640)	920 (3236)	800 (2810)	800 (2810)	595 (2095)	595 (2095)
Nominal size machine	tons (kW$_T$)	762 (2670)	667 (2350)	1,106 (3889)	1,013 (3557)	816 (2867)	550 (1934)	570 (2005)
Heat rejection	Btuh (kW$_T$)	18,000,000 (5230)	16,875,000 (4890)	20,700,000 (5997)	18,400,000 (5210)	18,000,000 (5230)	17,900,000 (5246)	17,900,000 (5246)
Cooling water ΔT	°F (°C)	10 (6)	10 (6)	10 (6)	20 (6)	10 (11)	10 (6)	10 (6)(6)
Total energy usage	kWh/yr	1,371,000	1,272,000	1,583,000	902,000	1,358,000	1,180,000	1,181,000
Total energy demand	kW	261	242	300	103	259	225	225
Absorption machine first cost	$	352,000	315,000	465,000	425,000	360,000	193,000	199,000

Note: The source heat stream is the exhaust gas from a gas turbine. The exhaust gas flowrate is 61,000 lb/hr (7.7 kg/s) at 960°F (516°C). The SI units are not a direct conversion; they were calculated separately. The amount of recovered heat is based on a gas-to-steam heat recovery system (or a gas-to-water in the case of the hot water unit) recovering steam at the listed quality. Compare the design cooling output for relative machine performance.

REQUIRED SYSTEM CHARACTERISTICS

TABLE 2-3
LiBr Absorption System Variations — Fixed Cooling Capacity

Description	Units	Base Steam Case	Double-Effect Increased Steam Press.	Double-Effect Hot Water	Double-Effect Increased Cooling Water ΔT	Double-Effect Decreased CHW Temp.	Single-Effect Steam	Single-Effect Decreased CHW Temp.
Leaving CHW temperature	°F (°C)	42 (6)	42 (6)	42 (6)	42 (6)	40 (4)	42 (6)	40 (4)
Recovered heat quality	psig steam (kPa)	115 (894)	130 (998)	80 (653)	115 (894)	115 (894)	12 (184)	12 (184)
Amount recovered heat	Btuh (kW$_T$)	8,692,000 (2563)	9,099,000 (2667)	8,087,000 (2370)	8,692,000 (2563)	8,692,000 (2563)	14,058,000 (4120)	14,058,000 (4120)
Design cooling output	tons (kW$_T$)	800 (2810)	800 (2810)	800 (2810)	800 (2810)	800 (2810)	800 (2810)	800 (2810)
Nominal size machine	tons (kW$_T$)	762 (2670)	712 (2502)	962 (3377)	1,013 (3557)	816 (2867)	740 (2603)	769 (2705)
Heat rejection	Btuh (kW$_T$)	18,000,000 (5230)	18,000,000 (5230)	18,000,000 (5230)	18,000,000 (5230)	18,000,000 (5230)	24,480,000 (7175)	24,480,000 (7175)
Cooling water ΔT	°F (°C)	10 (6)	10 (6)	10 (6)	20 (11)	10 (6)	10 (6)	10 (6)
Total energy usage	kWh/yr	1,371,000	1,362,000	1,377,000	902,000	1,379,000	1,584,000	1,581,000
Total energy demand	kW	261	260	261	103	261	179	179
Absorption machine first cost	$	320,000	356,000	404,000	425,000	342,000	261,000	271,000

Note: The design cooling output is constant at 800 tons (2810 kW$_T$). The amount of recovered heat and the absorption machine size are allowed to vary.

These tables are presented in order to illustrate the basic differences between single- and double-effect absorption machines and how varying the input conditions effects the output conditions.

2.5 COOLING ESTIMATES

If the quantity of recovered heat is known, the amount of cooling can be approximated using Equation 2-1.[1] This equation is based on the COP of the absorption machine:

$$q_{cooling} = COP_{ABS} \frac{q_{recovered}}{C_1} \qquad (2\text{-}1)$$

where
- $q_{cooling}$ = cooling output, tons (kW_T);
- COP_{ABS} = coefficient of performance of the absorption machine;
- $q_{recovered}$ = heat recovered from jacket/exhaust gas, Btu/h (kW_T); and
- C_1 = constant, 12,000 Btuh/ton (1 kW_T/kW_T).

2.6 SUMMARY

The information provided in this chapter allows for the identification of components required by absorption cooling systems. Table 2-4 summarizes the information presented in this chapter.

TABLE 2-4
Absorption System Requirements Summary

	Single-Effect LiBr Machine	Double-Effect LiBr Machine	Low-Temperature Ammonia Machine
Steam input, psig	5–15	65–145	0–200
(kPa)	(135–205)	(110–120)	(100–1480)
Steam input, °F	225–250	305–365	210–385
(°C)	(110–120)	(175–185)	(100–195)
Hot water or process fluid input, °F	240–300	310–400	210–385
(°C)	(115–150)	(155–205)	(100–195)
Cooling output, °F	40–100	40–80	−60 to 40
(°C)	(4–38)	(4–27)	(−51 to 4)
COP	0.6–0.7	0.9–1.2	0.1–0.8

Note: This table is based on manufacturers' data available in 1994.

---

[1] Equation 2-1 is for initial estimating only. In this Guide "C's" are for constants and "K's" are for variables found in tables or figures.

The values shown in Table 2-4 are recommended ranges for the economical application of absorption systems. The economic impact of absorption systems is contained in the next chapter.

REFERENCES

ASHRAE. 1992. *1992 ASHRAE handbook—HVAC systems and equipment.* Chapter 7, "Cogeneration Systems," p. 7.10. *ASHRAE Handbook—HVAC Systems and Equipment.* Atlanta, Georgia: American Society of Heating, Refrigerating and Air-Conditioning Engineers, Inc.

ASHRAE. 1993. *1993 ASHRAE handbook—Fundamentals.* Chapter 1, Thermodynamics and Refrigeration Cycles. Atlanta: American Society of Heating, Refrigerating and Air-Conditioning Engineers, Inc.

Boyen, J. L. 1980. *Thermal energy recovery*, 2d ed. New York: John Wiley & Sons, Inc.

Manufacturer. 1975. *Hermetic absorption liquid chillers.* Form 16JP-3P. Syracuse, NY: Carrier Corp.

Manufacturer. 1989. *Single stage absorption cold generator, 101 to 1660 tons.* Bulletin ABS-DS-1. LaCrosse, WI: The Trane Co.

Rohrer, W. M., Jr., and K. G. Kreider. 1979. *Industrial and institutional waste heat recovery.* Chapter 1, Sources and Uses of Waste Heat. Noyes Data Corporation.

Smith, W. P., Jr., R. C. Erickson, W. A. Liegois, and C. E. Dorgan. 1992. *Cogeneration technology.* Course text. Madison: College of Engineering, EPD Department, University of Wisconsin.

Vicatos, G., and J. Gryzagoridis. 1994. Unpublished research, Department of Mechanical Engineering. Cape Town: University of Cape Town, South Africa.

CHAPTER 3

Economics of Absorption Systems

This chapter analyzes the general costs associated with absorption systems. The costs covered in this chapter are for a basic system and include the absorption equipment, heat rejection equipment, and installation expenses.

The costs presented in this chapter produce a rough estimate of capital costs and operating costs for the initial feasibility analysis. A life-cycle cost analysis must be completed using detailed load profiles and construction costs to determine the economic feasibility of a system. The costs shown are obtained from nationwide averages, pricing books, and rules-of-thumb used in the industry.

The information from chapter 2 pertaining to the quality and quantity of recovered heat, cooling, and heat rejection requirements is used in this chapter to determine the first cost and operating costs of a system. Information from equipment manufacturers and computer programs is available to provide a more accurate estimate and should be utilized as the actual design progresses.

3.1 COSTS

An initial estimate of the costs must be accomplished in order to determine the economic feasibility of any project. Costs include equipment, installation, operating costs, and maintenance costs. With this information, different types of systems may be compared and the most economically feasible may then be chosen for further analysis as appropriate.

For this guide, it is assumed that the costs to install and operate the alternative cooling options, typically vapor-compression or direct-fired absorption systems, are known. This guide covers only the information required to determine the cost of an indirect-fired absorption system.

3.2 FIRST COSTS

The first cost for an absorption system includes the absorption machine, heat rejection equipment, heat recovery equipment, and auxiliary equipment. Costs associated with preparing the site or constructing a building are not included in this guide. This cost may be significantly different for different systems. The absorption system's physical size will be larger than for a vapor-compression system of a similar capacity. This increase in size can translate to larger buildings, moving equipment, and support systems. Also, the absorption machine cannot be dismantled to fit through a small opening. This would compromise the machine's hermetic integrity.

3.2.1 Absorption Machine

Determining the cost of an absorption machine can be a complex task. For the purposes of this guide, the cost of an absorption machine (with power and piping hookups) has been summarized by machine size in Table 3-1. The numbers in Table 3-1 are based on average actual project costs (Smith 1990; Dorgan 1994) and nationally available pricing books (Means 1994; Delta Research 1994; ARI 1993). An average of the costs listed in Table 3-1 can be used for initial estimation purposes.

TABLE 3-1
Absorption Machine Capital Costs

Nominal Size[*] tons (kW_T)	Lithium Bromide-Water Single-Effect $/ton ($/kW_T$)	Lithium Bromide-Water Double-Effect $/ton ($/kW_T$)	Aqueous Ammonia $/ton ($/kW_T$)
250 (900)	330–660 (95–190)	460–920 (130–260)	1,200 (340)
500 (1750)	220–490 (65–140)	310–690 (90–195)	1,200 (340)
1,000 (3500)	210–390 (60–110)	290–550 (80–155)	Custom[**]
1,500 (5300)	200–380 (55–110)	280–530 (80–150)	Custom[**]

[*] At ARI conditions (ARI 1992).
[**] AAR machines larger than 500 tons (1,760 kW_T) are custom designed and built by the manufacturer. No costs were available.

3.2.2 Heat Rejection Equipment

The primary heat rejection equipment used in an absorption system is a cooling tower. Use of other heat rejection options, such as boiler feed water preheating, can help reduce the cost of heat rejection equipment. Typically, a cooling tower in an absorption system has centrifugal or propeller fans. A centrifugal fan unit has higher capital and operating costs, is smaller, and generates less noise compared to a propeller fan unit.

At ARI conditions (78°F [26°C] WB, 95°F to 85°F [35°C to 29°C] cooling water), the costs for centrifugal fan cooling towers and propeller fan cooling towers are shown in Table 3-2. The cost for the cooling water pump and piping are included in the costs and is approximately $108/ton ($31/kW$_T$) (Means 1994; Manufacturers 1994a). These values are on a per-ton (kW$_T$) basis of the design cooling output of the absorption machine and include installation costs.

TABLE 3-2
Cooling Tower Costs for Absorption Systems (see notes)

Cooling Water Temperature Difference, °F (°C)	Type of Cooling Tower	Single-Effect Machine, $/ton ($/kW$_T$)	Double-Effect Machine, $/ton ($/kW$_T$)
10 (6)	Centrifugal	250 (70)	220 (63)
	Propeller	240 (68)	210 (60)
15 (8)	Centrifugal	220 (63)	200 (57)
	Propeller	215 (61)	195 (55)
20 (11)	Centrifugal	210 (60)	190 (54)
	Propeller	205 (58)	185 (53)

Note: This table is based on a per-ton (kW$_T$) basis of the design cooling (evaporator) output. It is important to realize that the numbers in this table are averages only. Even though a cooling tower for an absorption system is larger than a similarly sized vapor-compression system, the increase in cooling tower cost does not increase linearly. The incremental cost for a cooling tower (cost to the contractor) is typically $25 to $30 per condenser ton ($7 to $9 per condenser kW$_T$) (Rockenfeller 1994).

When the temperature difference of the water across a cooling tower is increased from 10°F (6°C) to 15°F (8°C), a smaller cooling tower and cooling water circulating pump are required. However, the penalty for using a smaller cooling tower is a decrease in absorption machine capacity due to a higher leaving cooling water temperature—100°F (38°C) versus 95°F (35°C). In some cases, the nominal size of the absorption machine must be increased (10% to 20%) to counteract this reduction in capacity.

If the absorption machine size is increased to maintain constant output, then the energy input to the machine will increase between 4% and 8%. This increase in energy use is due to the varying characteristics of the two machines at the respective operating points (Manufacturers 1994b).

During the initial stages of a project, the procedure developed in chapter 8 can be used to determine the amount of heat rejection for an absorption machine and approximate the cost from the above numbers. The tradeoffs between cooling tower and absorption machine sizing can also be approximated. However, a manufacturer should be contacted as the project progresses to determine detailed sizing parameters.

3.2.3 Heat Recovery Equipment

A significant piece of equipment in an absorption system is the heat recovery equipment. The type and cost of the heat recovery equipment will change depending on the properties of the source heat stream and the recovered heat stream. For example, the cost for the heat recovery equipment increases due to the requirement for corrosion-resistant materials.

Due to the wide range of applications, it is not feasible to list all of the costs individually. However, for heat recovery steam generators an initial estimate ranging from $15 to $40/lb/h ($33 to $88/kg/h) can be used (Smith 1994). A value of $25/lb/h ($55/kg/h) is recommended. The costs for liquid heat exchangers are much more variable and are not estimated in this guide.

3.2.4 Auxiliary Equipment

Auxiliary equipment primarily includes the absorption system electric equipment and pump motors, fan motors, and the water treatment system for the cooling tower. These costs are not estimated in this guide. For initial estimation purposes, the cost difference between absorption and vapor-compression systems can be assumed to be zero. For individual systems, capital savings may be available due to the different configurations and requirements of the systems.

3.3 OPERATING COSTS

After the capital costs for an absorption system are estimated, it is necessary to determine the operating costs. Operating costs include the electrical use of the absorption machine, the energy penalty associated with the heat recovery equipment, and the electrical use of the condenser water pump and the cooling tower fans. For initial estimates, it can be assumed that the energy use of the chilled-water pump does not vary by system type (absorption versus vapor compression).

3.3.1 Absorption Machine Energy Use

Electrical energy use for an absorption machine results from the solution/refrigerant pump(s). Energy use can be estimated through the use of Table 3-3. The energy use for an absorption machine can be estimated by using the appropriate value from Table 3-3 and Equation 3-1.

ECONOMICS OF ABSORPTION SYSTEMS

$$AM_{kWh} = K_1 \cdot AM_{cap} \cdot H_1 \qquad (3\text{-}1)$$

where

AM_{kWh}	=	absorption machine energy use, kWh/yr;
K_1	=	value from Table 3-3, kW/ton (kW/kW$_T$);
AM_{cap}	=	absorption machine nominal size, tons (kW$_T$); and
H_1	=	hours absorption machine operational per year, hours/yr.

TABLE 3-3
Absorption Machine Electrical Energy Use

Machine Type	Machine Size, tons (kW$_T$)	Electrical Use, kW/ton (kW/kW$_T$)	Recommended Value, kW/ton (kW/kW$_T$)
Single-Effect	100–200 (350–700)	0.014–0.055 (0.004–0.016)	0.030 (0.009)
	200–400 (700–1400)	0.011–0.028 (0.003–0.008)	0.022 (0.006)
	400–600 (1400–2100)	0.010–0.019 (0.003–0.005)	0.015 (0.004)
	600–1000 (2100–3500)	0.008–0.015 (0.002–0.004)	0.012 (0.003)
	1000–1600 (3500–5600)	0.006–0.017 (0.002–0.005)	0.012 (0.003)
Double-Effect	100–250 (350–900)	0.019–0.023 (0.005–0.007)	0.021 (0.006)
	250–450 (900–1600)	0.015–0.028 (0.004–0.008)	0.020 (0.006)
	450–800 (1600–2800)	0.011–0.027 (0.003–0.008)	0.018 (0.005)
	800–1500 (2800–5300)	0.009–0.026 (0.003–0.007)	0.015 (0.004)

Notes: The tons (kW$_T$) listed are for a nominal-sized machine at ARI conditions (44°F [7°C] chilled water, 85°F [29°C] condenser water, 12 psig [184 kPa] steam single-effect, and 115 psig [896 kPa] steam double-effect).

The kW/ton (kW/kW$_T$) ranges are a compilation of those of most major manufacturers. The recommended kW/ton (kW/kW$_T$) is a weighted average and is a good approximation for an average use. No values were available for AAR systems.

The electrical demand of the absorption machine can be estimated using Equation 3-2:

$$AM_{kW} = K_1 \cdot AM_{cap} \qquad (3\text{-}2)$$

where

AM_{kW}	=	absorption machine electrical demand, kW;
K_1	=	value from Table 3-3, kW/ton (kW/kW$_T$); and
AM_{cap}	=	absorption machine nominal capacity, tons (kW$_T$).

3.3.2 Energy Penalty of Heat Recovery Equipment

The process of recovering heat requires the expenditure of some energy. Some examples of the energy penalty include the following.

- Increased backpressure in the exhaust stack of a generator is due to the addition of a heat recovery boiler. The increased backpressure decreases the electrical output of the generator. This decrease in electrical output ranges from 2% to 8% (Smith et al. 1992).
- Additional pumping energy is required. This is due to the increased pressure loss in a liquid stream from the addition of the absorption machine's generator heat exchanger. Without the absorption machine, the liquid heat stream would go to the environment or to a cooling tower. On most systems, this is a negligible cost and can be assumed to be zero for initial feasibility estimates.

3.3.3 Cooling Water Pump and Cooling Tower Fan Energy Use

The cooling water pump and cooling tower fans are the major electrical energy users of an absorption cooling system. In optimizing performance and cost, the exact sizing and consequent energy use of these components can become complicated. The numbers in this section are averages and generalizations of current manufacturers' ratings.

At ARI conditions, the cooling water pump for a single-effect absorption system circulates approximately 3.6 gpm/ton (0.06 L/s·kW$_T$). A double-effect absorption system typically requires 4.0 gpm/ton (0.07 L/s·kW$_T$). A vapor-compression system requires 3 gpm/ton (0.05 L/s·kW$_T$) (ASHRAE 1992). It should be noted that these are different temperature changes in the cooling apparatus. Refer to section 1.4.3 and Table 3-4 for more information.

The size of the cooling water pump can be estimated using the data contained in Table 3-4. These data correspond to an absorption machine at full load, with the steam input to the absorption machine generator at 12 psig (184 kPa) for single-effect systems and 115 psig (896 kPa) for double-effect systems.

ECONOMICS OF ABSORPTION SYSTEMS

TABLE 3-4
Cooling Water Pump Sizing

Condenser Water Temperature Difference	Units	Single-Effect Machine	Double-Effect Machine
10°F (6°C)*	gpm/ton (L/s/kW$_T$)	6.0 (0.1)	4.5 (0.08)
	bhp/ton (kW/kW$_T$)	0.12 (0.025)	0.1 (0.021)
12°F (7°C)*	gpm/ton (L/s/kW$_T$)	...	4.0 (0.07)
15°F (8°C)	gpm/ton (L/s/kW$_T$)	4.1 (0.07)	3.0 (0.05)
	bhp/ton (kW/kW$_T$)	0.09 (0.019)	0.06 (0.013)
17°F (9°C)*	gpm/ton (L/s/kW$_T$)	3.6 (0.06)	...
20°F (11°C)	gpm/ton (L/s/kW$_T$)	3.1 (0.06)	2.3 (0.04)
	bhp/ton (kW/kW$_T$)	0.06 (0.013)	0.05 (0.011)
25°F (14°C)	gpm/ton (L/s/kW$_T$)	2.45 (0.04)	1.8 (0.03)
	bhp/ton (kW/kW$_T$)	0.05 (0.011)	0.04 (0.008)
30°F (17°C)	gpm/ton (L/s/kW$_T$)	2.05 (0.04)	...
	bhp/ton (kW/kW$_T$)	0.04 (0.008)	...
35°F (19°C)	gpm/ton (L/s/kW$_T$)	1.75 (0.03)	...
	bhp/ton (kW/kW$_T$)	0.04 (0.008)	...

* ARI rating conditions for vapor compression, double-effect and single-effect, respectively. See chapter 1, section 1.4.3, for more information.

Notes: The gpm/ton (L/s/kW$_T$) ratings are per design ton (kW$_T$) of cooling output of the absorption machine. The bhp/ton (kW/kW$_T$) ratings of the cooling water pump(s) are based on a system with a 60-foot (179-kPa) head and a motor that is approximately 80% efficient.

Numbers in this table are based on the average of manufacturers' published ratings for absorption machines and cooling water pumps. AAR systems will have requirements similar to those of a double-effect system. See chapter 10 for more details on AAR systems.

To estimate the electrical use of the condenser water pump(s), use Equation 3-3:

$$CWP_{kWh} = \frac{K_2 \cdot AM_{cap} \cdot H_1 \cdot C_2}{\eta_{CWP}} \quad (3\text{-}3)$$

where

CWP_{kWh} = cooling water pump electrical use, kWh/yr;
K_2 = value from Table 3-4, bhp/ton (kW/kW$_T$);
AM_{cap} = absorption machine nominal size, tons (kW$_T$);
H_1 = hours absorption machine is operational per year, hours/yr;

C_2 = conversion factor, 0.7457 kW/hp (1 kW/kW$_T$); and
η_{CWP} = cooling water pump motor efficiency, %.

The electrical demand for the cooling water pump can be determined using Equation 3-2 and substituting $K_2 \cdot C_2 / \eta_{CWP}$ for K_1.

Estimating the energy use for the cooling tower fans is similar to the above procedure and can be accomplished using Table 3-5.

TABLE 3-5
Cooling Tower Fan Sizing

Condenser Water Temperature Difference	Type of Fan	Single-Effect Machine bhp/ton (kW/kW$_T$)	Double-Effect Machine bhp/ton (kW/kW$_T$)
10°F (6°C)	Propeller	0.13 (0.028)	0.11 (0.023)
	Centrifugal	0.27 (0.057)	0.23 (0.049)
15°F (8°C)	Propeller	0.11 (0.023)	0.09 (0.019)
	Centrifugal	0.23 (0.049)	0.19 (0.040)
20°F (11°C)	Propeller	0.09 (0.019)	0.08 (0.017)
	Centrifugal	0.19 (0.040)	0.17 (0.036)

Notes: The bhp/ton (kW/kW$_T$) ratings are per design ton (kW$_T$) of cooling output of the absorption machine and are based on a chilled-water temperature of 44°F (7°C) and an input of 12 psig (184 kPa) steam for a single-effect machine and 115 psig (894 kPa) steam for a double-effect machine.

Numbers in this table are based on the average of manufacturers' published ratings for absorption machines and cooling towers.

The electrical use of the cooling tower fans can be estimated by using Equation 3-4:

$$\text{CTF}_{kWh} = \frac{K_3 \cdot AM_{cap} \cdot H_1 \cdot C_2 \cdot K_4}{\eta_{CTF}} \quad (3\text{-}4)$$

where
CTF_{kWh} = cooling tower fan electrical use (kWh/yr);
K_3 = value from Table 3-5, bhp/ton (kW/kW$_T$);
AM_{cap} = absorption machine nominal size, tons (kW$_T$);
H_1 = hours absorption machine is operational per year, hours/yr;
C_2 = conversion factor, 0.7457 kW/hp (1 kW/kW$_T$);
K_4 = conversion factor, cooling tower fan partial use, %; and
η_{CTF} = cooling tower fan motor efficiency, %.

The variable K_4 is a part-use factor for the cooling tower fans. A part-use factor is required for a cooling tower since the cooling tower fan does not operate continuously, but operates intermittently to maintain the leaving cooling water temperature. The amount of time the fans operate will vary, depending on the local weather conditions. If known, the part-use factor for the specific location should be used. Otherwise, a value of 0.4 can be used for initial energy use calculations. For the later stages of design, a more detailed estimate of energy use will need to be accomplished using an hourly or bin modeling program.

The electrical demand for the cooling tower fans can be estimated using Equation 3-2 and substituting $C_2 \cdot K_3 / \eta_{CTF}$ for K_1.

3.4 MAINTENANCE COSTS

The final cost to be estimated for an absorption cooling system is the cost of maintaining the system. Maintenance costs are highly variable and depend on local labor rates, age of the absorption machine, length of time the absorption machine is operated, and the amount of preventive maintenance that is performed on the system. The maintenance recommendations and requirements of the machine manufacturer should be accomplished as a minimum.

As an absorption machine ages and daily operating time increases, the maintenance costs increase. Regardless of the system, the maintenance costs will vary depending on the building's geographical location (urban or rural) and the expertise of the maintainers. At the current time, there is no consensus of the variations in maintenance costs between electrical and absorption cooling systems. Maintenance costs cited in various studies and surveys show that absorption machines cost from 0.6 to 1.25 times more than electric chillers (Gilbert and Associates 1993; Pawelski 1994; Reimann 1994). Maintenance costs for AAR systems were unavailable.

These are rough estimates for comparison purposes only and should not be used for budgeting purposes. Quotes from local maintenance firms should be obtained for the completion of a budget. Qualified maintenance firms can be found in most large cities. One of the best sources for locating a suitable contractor is to contact the local or corporate offices of the equipment manufacturers.

3.5 SUMMARY OF COSTS

Table 3-6 provides a form to summarize the costs to purchase, operate, and maintain an absorption cooling system. As previously discussed, the values included in this chapter are for initial estimating purposes in doing screening or feasibility analysis. A detailed life-cycle cost analysis of the components must be completed to determine the actual costs of any system over the lifetime of the equipment.

APPLICATION GUIDE FOR ABSORPTION COOLING/REFRIGERATION USING RECOVERED HEAT

TABLE 3-6
Economic Summary

FIRST COSTS
- Absorption
  Machine capacity (AM_{CAP}) _____ tons (kW_T)
  Capital cost per ton (kW_T), Table 3-1 _____ $/ton ($/kW_T)
  Total cost = $AM_{CAP} \cdot$ $/ton ($kW_T$) $ _____

- Heat Rejection Equipment
  Cost per ton (kW_T), Table 3-2 _____ $/ton ($/kW_T)
  Total cost = $AM_{CAP} \cdot$ $/ton ($kW_T$) $ _____

- Heat Recovery Equipment
  $25/lb/h$\cdot AM_{CAP}\cdot$steam rate/cooling unit $ _____

- Total First Cost $ _____

OPERATING COSTS
- Absorption Machine
  kW per ton, (kW/kW_T), Table 3-3, K_1 _____ kW/ton (kW/kW_T)
  Hours operational (system) per year, H_1 _____ hours/yr
  Total demand = $AM_{CAP} \cdot K_1$ _____ kW _____ kWh/yr
  Total use = $K_1 \cdot AM_{CAP} \cdot H_1$
- Cooling Water Pump
  Fan horsepower per ton (kW/kW_T), _____ hp/ton (kW/kW_T)
  Table 3-4, K_2
  Hours operational (system) per year, H_1
  Total demand = $K_2 \cdot 0.7457 \cdot AM_{CAP}/M_{CAP}$ _____ kW
  Total use = $K_2 \cdot 0.7457 \cdot AM_{CAP} \cdot H_1/M_{CAP}$ _____ kWh/yr

- Cooling Tower Fans
  Pump horsepower per ton, (kW/kW_T), _____ hp/ton (kW/kW_T)
  Table 3-5, K_3
  Hours operational (system) per year, H_1 _____ hours/year
  Total demand = $K_3 \cdot 0.7457 \cdot AM_{CAP}/M_{CTF}$ _____ kW
  Total use = $K_3 \cdot 0.7457 \cdot AM_{CAP} \cdot H_1/M_{CTF}$ _____ kWh/yr

Note: The 0.7457 should be replaced with 1.0 if using SI units.

REFERENCES

ARI. 1992. *Standard 560, Absorption water chilling and water heating packages.* Arlington, VA: Air-Conditioning and Refrigeration Institute.

ARI. 1993. Annual sales figures. Arlington, VA: Air-Conditioning and Refrigeration Institute.

ASHRAE. 1992. *1992 ASHRAE handbook—HVAC systems and equipment,* chapter 37, p. 37.2. Atlanta: American Society of Heating, Refrigerating and Air-Conditioning Engineers, Inc.

Delta Research Corporation. 1994. Tri-services automated cost engineering system (TRACES) cost estimating analysis, system design. Niceville, FL: Delta Research Corp.

Dorgan Associates, Inc. 1994. Survey of HVAC contractors across the United States. Madison, WI: Dorgan Associates, Inc.

Gilbert and Associates. 1993. *CFCs and electric chillers, selection of large capacity water chillers in the 1990s (revision 1).* EPRI TR-100537, R1. Palo Alto, CA: Gilbert and Associates.

Manufacturers. 1994a. Personal communications with BAC and Marley (fax).

Manufacturers. 1994b. Personal communication with Carrier, McQuay, Trane, and York (phone).

Means, R.S. 1993. *Means mechanical cost data 1994,* 17th annual edition. Kingston, MA: R.S. Means.

Pawelski, M. 1994. Personal communication (meeting).

Reimann, B. 1994. Personal communication (letter).

Rockenfeller, U. 1994. Personal communication (meeting and phone). Information is based on data from the Cooling Tower Institute.

Smith, W.P., Jr. 1990. *Absorption refrigeration: Performance at the USA sites.* Presented at Annual RE Technology Exchange-BASF, June.

Smith, W.P., Jr. 1994. Personal communications (letter and phone).

Smith, W.P., Jr., R.C. Erickson, W.A. Liegois, and C.E. Dorgan. 1992. *Cogeneration technology.* Course text, College of Engineering, EPD Department. Madison: University of Wisconsin.

CHAPTER 4
Heat Recovery

This chapter provides a general overview of the available methods for recovering heat, the advantages and disadvantages of various systems, and general guidance on selecting the most useful heat recovery method. Heat recovery methods described in this guide are restricted to technology available for use with absorption cooling machines. This chapter is not intended to provide all information required to design a heat recovery system.

4.1   METHODS OF HEAT RECOVERY

Recovering heat for use in an absorption machine can be accomplished either directly or indirectly, depending on the chemical composition of the source heat stream. Direct use of the source heat stream saves on initial capital investment, as it is unnecessary to purchase heat recovery equipment. An added benefit is that there are no heat exchanger efficiency losses. Conversely, operating efficiency may drop if fouling agents in the source heat stream are allowed to build up in the absorption machine.

Use of a heat exchanger to generate steam or hot water can reduce the problems associated with the direct use of the source heat stream. Total system initial capital investment is higher due to the cost of the heat exchange equipment. There may be an increase in maintenance costs for the additional equipment. Also, some energy is lost at the heat exchanger. The peak available temperature will be reduced, depending on the heat exchanger effectiveness.

4.1.1 Steam

The most widely used method of recovering heat for absorption cooling is through the use of a heat recovery steam generator (Boyen 1980).

4.1.1.1 Steam Generation for Single-Effect Absorption Machines

Steam at pressures of less than 15 psig (200 kPa) is the most common heat source for single-effect absorption machines (Boyen 1980). Generating this steam by using recovered heat is common practice. Heat sources that can generate steam for single-effect absorption machines include reciprocating engine cooling jackets and exhausts, gas turbine exhausts, incinerators' effluent gases, steam turbine exhausts, and industrial process.

4.1.1.2 Steam Generation for Double-Effect and Ammonia-Absorption Machines

Double-effect and low-temperature absorption machines typically use steam at pressures between 65 and 145 psig (550 and 110 kPa). Steam at these pressures is advantageous when it must be transported long distances, as in a campus situation or where it is required for industrial processes that need high-temperature heat sources.

Generating steam at these pressures from a source heat stream is most often done with high-temperature exhaust gases from gas turbines or incinerators (Boyen 1980). Heat recovery boilers are available to suit most situations, but supplemental firing may be required to match cooling loads and steam generation.

While steam pressures of more than 145 psig (1,100 kPa) are not typically used for air-conditioning (Boyen 1980), pressures up to 200 psig (1,480 kPa) are used for refrigeration processes in aqueous-ammonia units (Shepherd 1994). Often, when steam generated at pressures of more than 145 psig (1,100 kPa) is used for power generation or other processes, lower pressure steam or condensate can be recovered from these operations and be used for absorption cooling.

4.1.2 Hot Water

While steam is the most popular heat transfer fluid, medium-temperature high-pressure water (230°F to 300°F [110°C to 150°C]) is often used as well. Hot water systems are typically used for heating in buildings where process steam is not needed. Hot water systems can be installed and operated at a lower cost than steam systems because less equipment and less maintenance are required. The use of hot water as a heat source will affect the sizing of the absorption machine.

High-pressure hot water can be generated by the same techniques used to generate low-pressure steam. One typical source is gas turbine exhaust.

4.2 HEAT RECOVERY EQUIPMENT

There are many different ways to generate the steam or hot water needed for the absorption process. While it may be possible to use the recovered heat stream directly, a variety of heat exchangers are available to recover heat to make steam and hot water. The

HEAT RECOVERY

following heat recovery equipment is available from a wide range of manufacturers and can be used alone or in combination to produce the required steam pressures or water temperatures:

- shell-and-tube heat exchangers (liquid or steam to hot water),
- exhaust gas boilers (exhaust gas to steam or hot water), and
- steam separators.

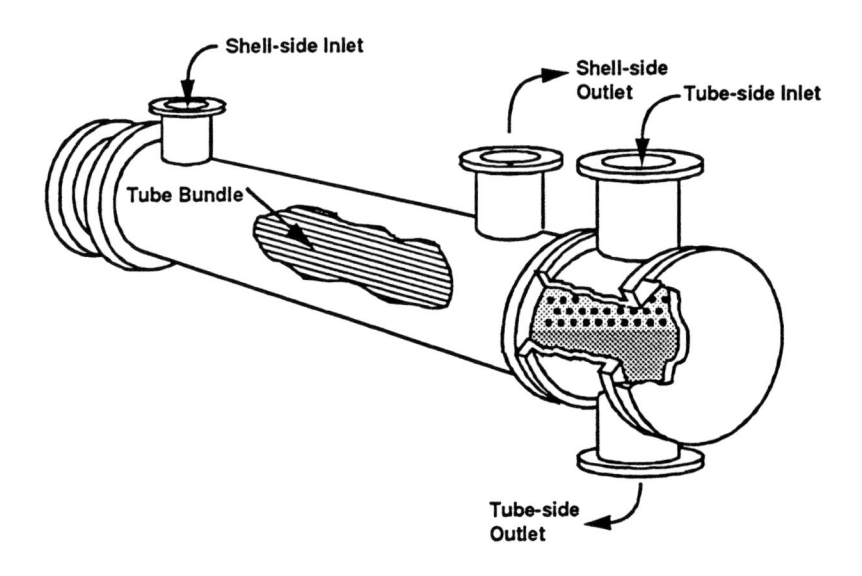

4.2.1 Shell-and-Tube Heat Exchangers (Liquid or Steam to Hot Water)

Shell-and-tube heat exchangers are commonly used to transfer heat from one liquid to another or to transfer heat from steam to hot water. A shell-and-tube heat exchanger is shown in Figure 4-1.

Figure 4-1 Shell and Tube Heat Exchanger

To promote heat transfer, the most viscous fluid is usually passed on the outside of the tubes, increasing turbulence. However, if one fluid tends to foul the tubes more easily than the other, that fluid should be passed on the inside of the tubes to ease cleaning (Manufacturer 1993).

Shell-and-tube heat exchangers are widely available and can be constructed using stainless steel, cupro-nickel, or any of a variety of materials that resist corrosion.

4.2.2 Exhaust Gas Boilers (Exhaust Gas to Steam or Hot Water)

Exhaust gases from gas turbines or stack gases from industrial processes can be used as heat sources to generate either steam or hot water for use in an absorption machine. Exhaust gas boilers can be constructed as either water-tube or fire-tube units and may also combine exhaust gas silencing. A water-tube unit is shown in Figure 4-2 and a fire-tube unit is shown in Figure 4-3.

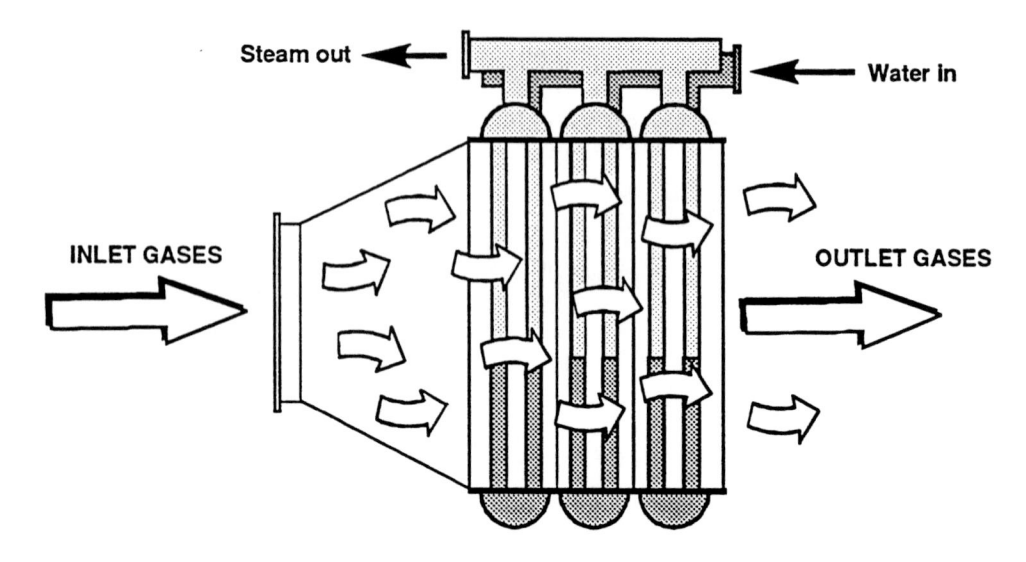

Figure 4-2 Water-Tube Boiler Using Exhaust Gases

Water-tube and fire-tube units each have advantages. Water-tube recovery systems typically provide better silencing of exhaust gases, have faster start-up times, and have high energy recovery at low cost. Fire-tube units are long lasting and have simple piping arrangements (Manufacturer 1984).

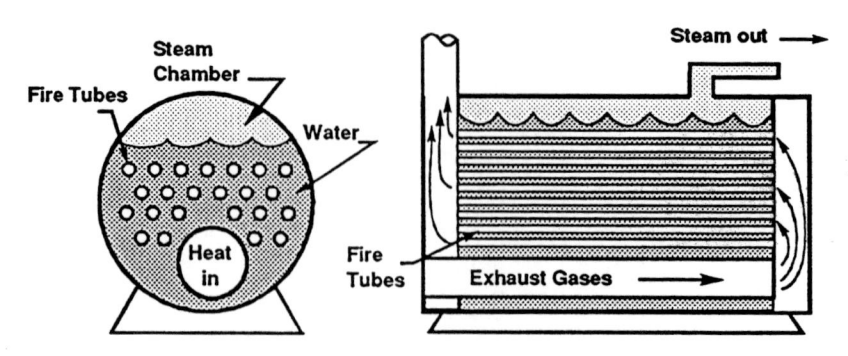

Figure 4-3 Fire Tube Exhaust Gas Boiler

HEAT RECOVERY

4.2.3 Steam Separators

Steam separators are used to recover high-quality steam from steam-water mixtures. The steam-water mixture is most often from an engine cooling water system. Steam separators are frequently combined with exhaust gas heat recovery silencers. A steam separator is shown in Figure 4-4.

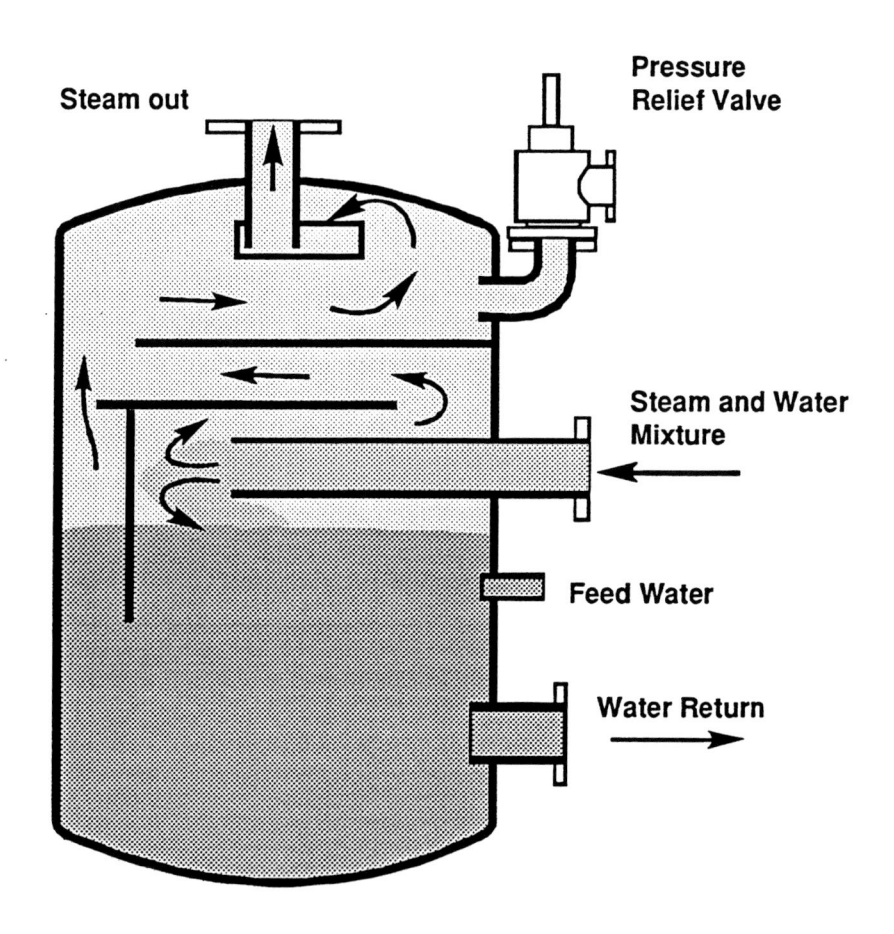

Figure 4-4 Steam Separator

Steam separators work by the following process. A steam-water mixture enters the separator and strikes the impact plate, causing initial separation of steam from water. The steam travels upward toward the steam outlet through a series of baffles, removing any remaining water. Water, including the water from the baffled section of the separator, cascades to the storage area in the bottom of the separator. The water is returned to the source, where it provides additional heating (Manufacturer 1990). Centrifugal steam separators are also available.

4.3 COGENERATION

Use of the rejected heat from cogeneration systems for absorption cooling is a common application and is currently very popular. By ASHRAE definition, *cogeneration* is the sequential production of two forms of useful energy from the same fuel source. Cogeneration typically involves the production of electricity and heat recovery. The heat recovered is used for water or space heating, absorption cooling, or other process applications (ASHRAE 1992a).

Many issues are involved in the decision to install a cogeneration system. Fuel and electricity costs, electric and steam load, as well as potential hours of use are all very important. Typically, a cogeneration system is economical if coincident electric power and heating/cooling needs are common and large. The following criteria can be used as a guide to determine where cogeneration is a viable option (Smith *et al*. 1992):

- low fuel cost (< $3.00/MBtu [$2.84/GJ]),
- high electric costs (> $0.05/kWh),
- used most of the hours in a year (> 6,000),
- high electric load (> 1000 kW; load factor > 70%),
- high steam requirements (> 10 MBtu/h [2,900 kW_T]; load factor > 70%),
- low steam pressure requirements at process (< 50 psig [440 kPa]), and
- steam pressure at boilers (> 200 psig [1,480 kPa], potential to use steam turbines)

These criteria are only an initial guide. Each system must be treated on an individual basis and may prove to be very economical without meeting many of the above criteria.

Several common types of cogeneration systems are described below. All can be used with absorption cooling.

4.3.1 Gas Turbines

Gas turbine cogeneration systems have several advantages over other types of cogeneration systems. The larger gas turbines are very reliable and often require less maintenance under continuous operation. Gas turbine systems also have high heat recovery capability, require no cooling water, have clean exhaust, and are attractive when all or most of the heat recovered can be used all the time (ASHRAE 1992b).

Exhaust gases from a gas turbine range from 800°F to 1000°F (430°C to 530°C). When used as part of a total energy system, the thermal efficiency of a gas turbine system can exceed 90% (ASHRAE 1992b). Typical gas turbine exhaust has between 10,000 and 17,000 Btuh/kW (2.93 and 4.98 kJ/kW·s) of generating capacity (Manufacturer 1994).

Exhaust gases also contain between 15% and 18% oxygen (Manufacturer 1983). This allows for supplemental firing of the exhaust gas stream through the use of an exhaust gas boiler and an increase in the amount of recovered heat.

Heat exchangers are available specifically for heat recovery from gas turbine exhaust. These can be used to generate either steam or hot water.

4.3.2 Steam Turbines

Steam turbines can be used in a variety of applications. They can be the prime movers in a cogeneration system or drive centrifugal compressors and chilled-water pumps for air conditioning (ASHRAE 1992b).

Recovering heat from steam turbines for use in absorption systems is done using a backpressure steam turbine. The turbine is designed to exhaust steam at the desired pressure required at the absorption machine generator. Automatic-extraction steam turbines are available that allow regulation of exiting steam pressure. Typical steam turbines operate using steam between 100 and 250 psig (790 and 1,830 kPa) and can extract steam up to 50 psig (450 kPa).

4.3.3 Reciprocating Engines

Reciprocating engines are often used to generate electricity in cogeneration systems. Engine-cooling jackets and exhaust heat boilers/mufflers are used to recover the generated heat. Heat can be recovered in the form of steam or hot water.

Reciprocating engine-cooling jackets are commercially available to suit almost any engine size (Boyen 1980). Steam pressure available from this type of system is limited to 15 psig (200 kPa), which is ideal for single-effect absorption cooling. Hot water units are also available and are well suited for absorption cooling.

When designing a heat recovery system for a reciprocating engine, it is important to remember that the primary function of the heat recovery equipment is to cool the engine. Secondary functions are muffling engine exhaust and recovering heat. Heat recovery boilers/mufflers can recover up to 40% of the heat generated by an engine (Manufacturer 1992).

Since demand for the recovered heat is not always constant, engine heat recovery systems can be designed with backup radiators or load-balancing condensers to remove heat not needed by the absorption system. This heat energy is stored for later use. Thermal energy storage can also be used to store excess cooling during periods having low cooling demand.

Reciprocating engine-cooling jackets that can provide low-pressure steam or water temperatures between 180°F and 250°F (82°C and 121°C) typically use a shell-and-tube heat exchanger to transfer the rejected engine heat to a secondary water circuit. Steam systems use additional equipment consisting of a flash boiler to convert the hot water to steam or can use natural (ebullient) circulation, which requires no pumps. An exhaust heat boiler/muffler may be used with this configuration to recover additional heat.

A forced-circulation reciprocating engine-cooling jacket is shown in Figure 4-5. Steam is generated in the flash boiler due to the pressure differential between the engine-cooling water outlet and the boiler. Steam is delivered to the load, and the remaining water is circulated back to the engine. Systems of this type typically operate at steam pressures up to 8 psig (155 kPa), which is within the acceptable range for single-effect absorption cooling.

Ebullient cooling of an engine involves natural circulation of the jacket cooling water. These systems are very simple, have low operating and maintenance costs, and provide steam at pressures well suited for absorption cooling—11.6 to 15 psig (180 to 200 kPa).

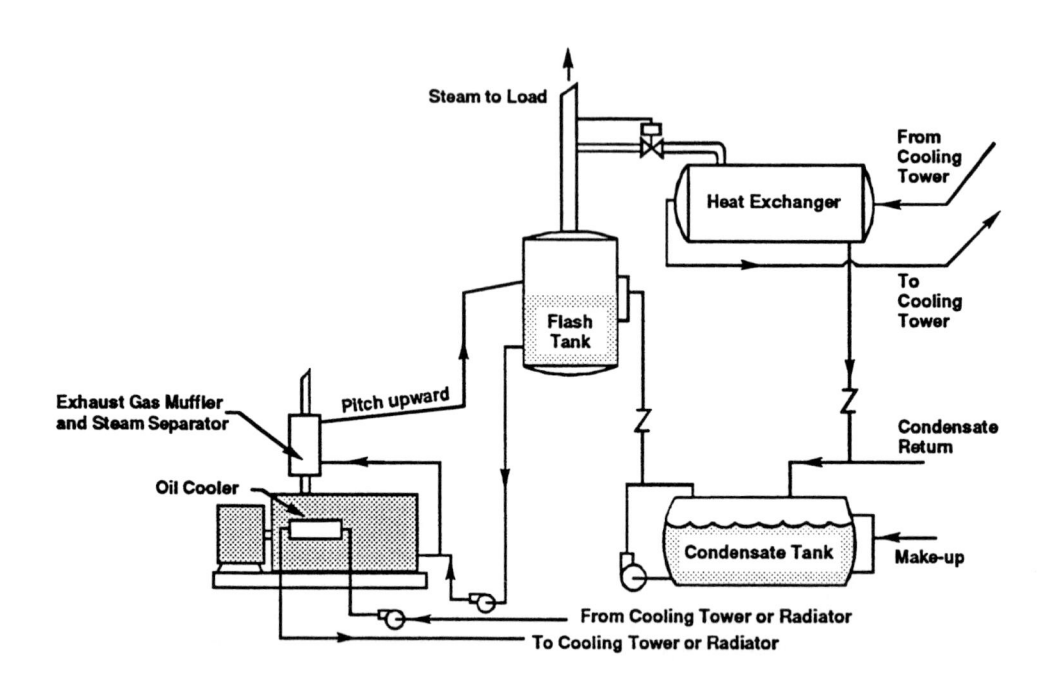

Figure 4-5 Forced Circulation Reciprocating Engine Cooling Jacket

REFERENCES

ASHRAE. 1992a. *1992 ASHRAE handbook—HVAC systems and equipment*, chapter 7. Atlanta: American Society of Heating, Refrigerating and Air-Conditioning Engineers, Inc.

ASHRAE. 1992b. *1992 ASHRAE handbook—HVAC systems and equipment*, chapter 41. Atlanta: American Societey of Heating, Refrigerating and Air-Conditioning Engineers, Inc.

Boyen, J.L. 1980. *Thermal energy recovery*, 2d ed. New York: John Wiley & Sons.

Manufacturer. 1983. *Cogeneration systems*. T75J/887/20M. San Diego, CA: Solar Turbines, Inc.

Manufacturer. 1984. *On-site power generation handbook.* Catalog No. LEBX4457. Milwaukee, WI: Caterpillar Engine Division.

Manufacturer. 1990. *Installation, operating and maintenance instructions Maxim TRP heat recovery silencers.* Bulletin 143 90TRP. Shreveport, LA: Beaird Industries, Inc.

Manufacturer. 1992. *Engine performance, G3400 gas engine.* Catalog No. LEBQ2024. Milwaukee, WI: Caterpillar Engine Division.

Manufacturer. 1993. *Cogeneration.* File No. 3019E 2.5M-4/93. Coltec Industries, Fairbanks Morse Engine Division.

Manufacturer. 1994. *U.S. Turbines catalog.*

Shepherd, J.J. 1994. Personal communication (letter and phone).

Smith, W.P., Jr., R.C. Erickson, W.A. Liegois, and C.E. Dorgan. 1992. *Cogeneration technology.* Course text, College of Engineering, EPD Department. Madison: University of Wisconsin.

CHAPTER 5

The Absorption Process

Chapter 5 describes the absorption process in terms parallel to those of the vapor-compression cycle. A thorough understanding of the absorption cycle is useful prior to discussing application requirements.

Chapter 5 is divided into three parts. The first is a brief review of the vapor-compression cycle, which establishes a knowledge base. The second is a description of the absorption process, including the less familiar concepts exclusive to the absorption process. The third is a description of the absorption fluid mixtures and their properties.

5.1 THE VAPOR-COMPRESSION CYCLE

This section is a review of the vapor-compression cycle, which is the system most often applied for air-conditioning and refrigeration applications. The basic vapor-compression cycle is used to illustrate the refrigeration process, since it is generally familiar to many engineers. The basic cycle is essentially the same regardless of the compressor type (reciprocating, screw, scroll, or centrifugal) or refrigerant (chlorofluorocarbon [CFC] or non-CFC) used.

Figure 5-1 shows a simplified schematic of the equipment layout for a typical vapor-compression cycle. Each component of the vapor-compression cycle—the compressor, condenser, evaporator, and expansion valve—is separated by piping.

The vapor-compression cycle can be graphically represented on a pressure-enthalpy (P-H) diagram. A P-H diagram displays the complete state of the refrigerant, including pressure, enthalpy, entropy, temperature, and phase (liquid, vapor, etc.). A generic P-H diagram is shown in Figure 5-2.

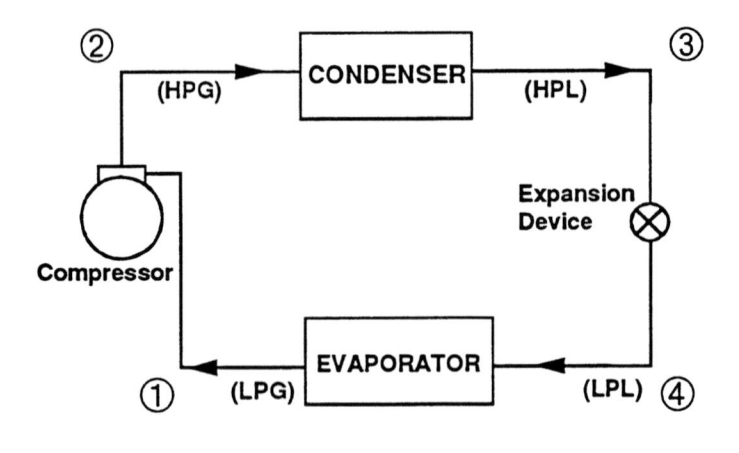

APPLICATION GUIDE FOR ABSORPTION COOLING/REFRIGERATION USING RECOVERED HEAT

Key:
HPG – high pressure gas
LPG – low pressure gas
HPL – high pressure liquid
LPL – low pressure liquid

Figure 5-1 Vapor Compression Cycle

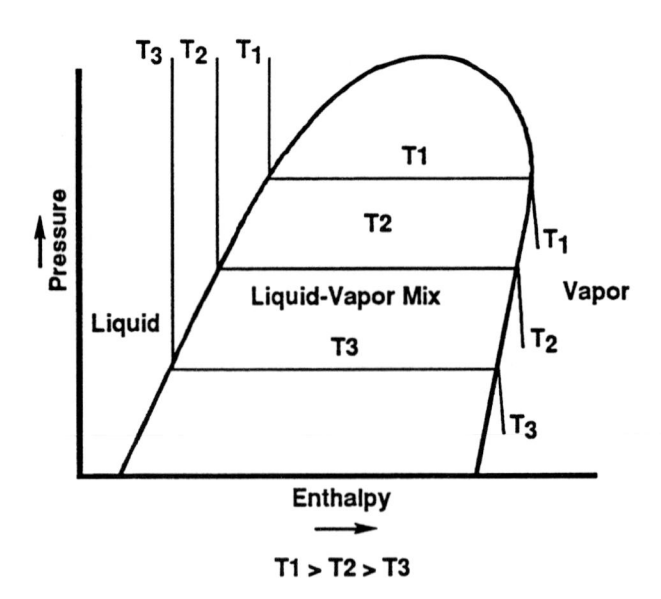

$T_1 > T_2 > T_3$

Figure 5-2 Generic P-H Diagram

48

Figure 5-3 illustrates the vapor-compression cycle on a P-H diagram for the equipment shown in Figure 5-1.

Path 1→2: At point 1, the refrigerant is a low-pressure vapor. The compressor increases the pressure of the vapor to point 2, with a related temperature increase.

Path 2→3: At point 2, the refrigerant is a high-pressure, high-temperature vapor. The pressure must have a corresponding saturated condensing temperature (SCT) higher than the condensing media. From point 2 to point 3, the vapor passes through a condenser where heat is removed, condensing the vapor to a subcooled liquid.

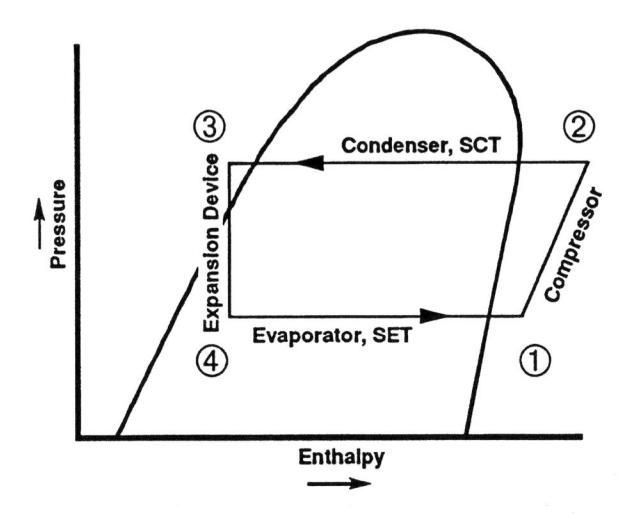

Figure 5-3 P-H Diagram of the Vapor Compression Cycle

Path 3→4: From point 3 to point 4, the liquid refrigerant is throttled through an expansion valve or other similar device, where some of the liquid flashes to vapor, forming a low-pressure, low-temperature two-phase mixture. The expansion device (valve) serves as the metering control to match the refrigeration load.

Path 4→1: From point 4 to point 1, the two-phase mixture passes through an evaporator. The refrigerant liquid-vapor mixture flashes to vapor by transferring heat from the chilled-water loop. The water passing through the evaporator is cooled and is circulated to the load. The refrigerant exiting the evaporator is a low-pressure vapor. The pressure must have a corresponding saturated evaporating temperature (SET) lower than the cooled media.

5.2   THE ABSORPTION CYCLE

The key difference between an absorption cycle and a vapor-compression cycle is the process by which the low-pressure refrigerant is transformed to a high-pressure vapor, path

1→2. In a vapor-compression cycle, the low-pressure vapor is mechanically compressed. In an absorption cycle, the low-pressure vapor is absorbed into a solution at low pressure, pumped to a high pressure, and then heated to produce a high-pressure vapor. The temperature differences between SET and SCT are similar for absorption and vapor-compression systems.

Figure 5-4 shows a schematic of an absorption cycle. The refrigerant cycle (path 2-3-4-1) is identical in both the vapor-compression cycle and the absorption cycle. However, the compressor function (path 1-2), is handled differently by each technology. In vapor compression, a mechanical compressor completes the compressor function. With absorption technology, the compressor function is replaced with a solution circuit and a pump. The compressor function in an absorption cycle is commonly referred to as a thermal compressor, thermochemical compressor, or a chemical compressor. In this guide, the terms *compressor function* and *thermal compressor* are used.

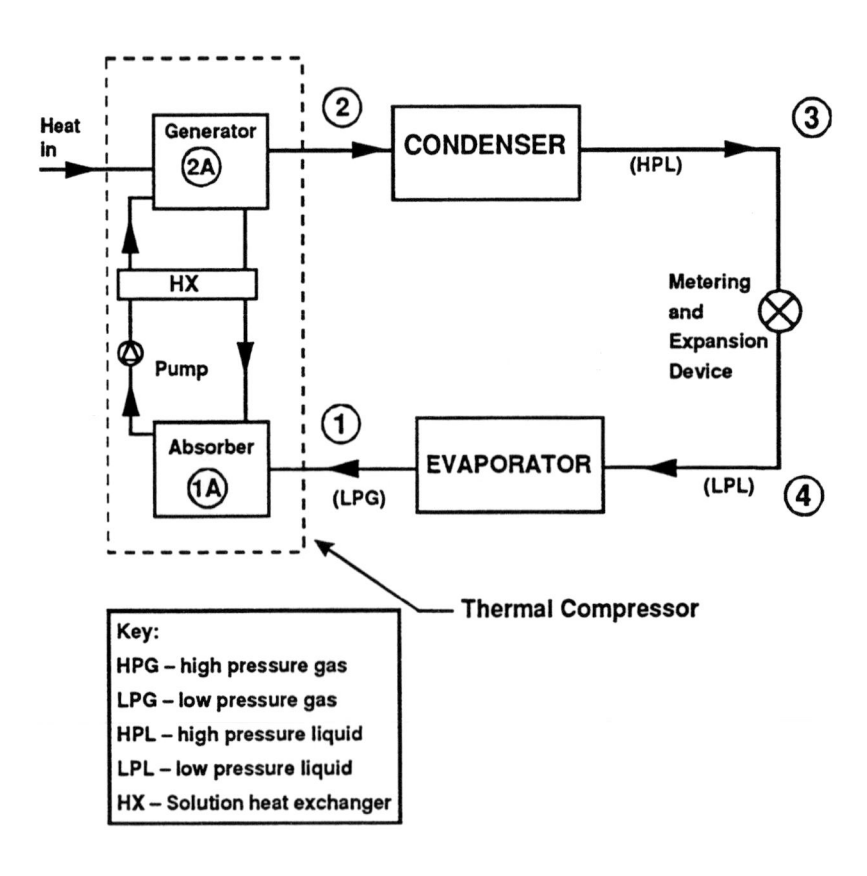

Figure 5-4 Absorption-Refrigeration Cycle

As shown in Figure 5-5, the P-H diagram of the refrigerant portion of an absorption cycle is similar to the P-H diagram of the vapor-compression cycle shown in Figure 5-4.

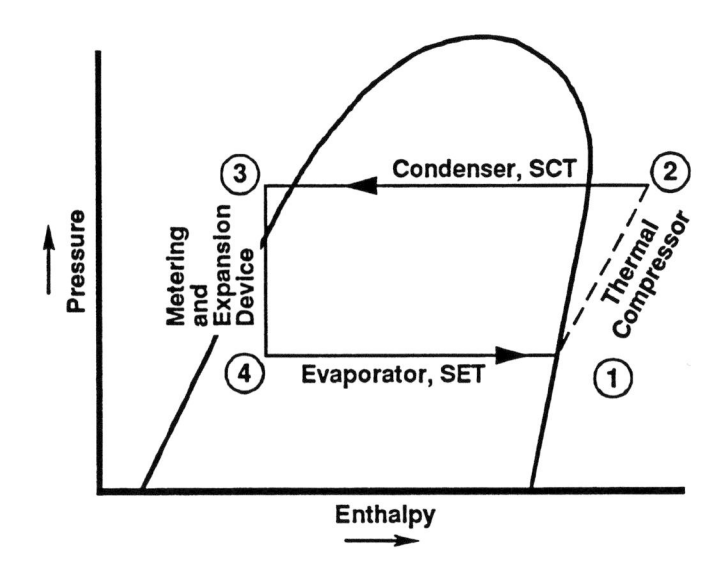

Figure 5-5 P-H Diagram of the Absorption Cycle (refrigerant only)

A description of the absorption cycle process shown in the P-H diagram in Figure 5-5 follows. All points referred to are the same in Figures 5-4 and 5-5.

Path 1→2: At point 1, the low-pressure, low-temperature refrigerant vapor enters the thermal compressor and is absorbed into a solution at point 1A. The combination of a pump and the addition of heat changes the state of the absorbent-refrigerant mixture from 1A to 2A. At point 2A, the refrigerant vapor is separated from the absorbent solution by heating (boiling), which generates a high-temperature, higher pressure refrigerant gas condition at point 2.

Path 2→3→4→1: The refrigerant gas completes the refrigeration cycle, which is identical to the vapor-compression cycle.

5.3   ABSORBENT-REFRIGERANT SOLUTIONS

In an absorption system, a mixture of a refrigerant and an absorbent form the solution used in the absorption cycle. Currently, two absorbent-refrigerant mixtures are widely employed. One is a lithium bromide (LiBr)-water mixture, also known as an aqueous LiBr mixture. The other is a water (aqueous) ammonia mixture. There are absorption cycles that use other solution combinations and refrigerants. Refer to chapter 11 for additional information.

5.3.1 Types of Absorption Systems

There are two primary absorption systems currently commercially available. These are LiBr-water systems and aqueous ammonia absorption refrigeration (AAR) systems. In a LiBr-water absorption system, the lithium bromide, a salt, is the absorbent and the water is the refrigerant. LiBr-water systems are the most commonly used absorption system, especially for commercial cooling.

In an AAR system, the water is the absorbent and the ammonia is the refrigerant. Ammonia systems are typically used when low-temperature cooling or freezing is required.

5.3.2 Solution Concentration Levels

The concentration of the solution changes throughout the absorption cycle. The amount of refrigerant in relation to the absorbent indicates the concentration level of the solution.

The naming convention for the solution concentration levels varies between the LiBr-water and aqueous ammonia systems. This frequently causes confusion with practioners. A discussion of the various terms used by ASHRAE (ASHRAE 1988, 1994) and the industry to describe solution concentration levels follows.

Weak Absorbent: The solution has a low capacity for absorbing refrigerant. The solution contains a high level of refrigerant in relation to the maximum that can be absorbed at a given temperature and proportionally lower levels of absorbent.

For LiBr systems, a weak absorbent is referred to as a *dilute solution*, since the proportion of absorbent to refrigerant in the solution is low (dilute).

For aqueous ammonia systems, a weak absorbent is referred to as a *strong aqua*, since the proportion of refrigerant to absorbent in the solution is high (strong).

Strong Absorbent: The solution has a high capacity for absorbing refrigerant at a given reference temperature. The solution contains a low level of refrigerant and proportionally high levels of absorbent.

For LiBr systems, a strong absorbent is referred to as a *concentrated solution*, since the proportion of absorbent to refrigerant in the solution is high (concentrated).

For aqueous ammonia systems, a strong absorbent is referred to as a *weak aqua*, since the proportion of refrigerant to absorbent in the solution is low (weak).

The weak/strong absorbent is ASHRAE terminology. The dilute/concentrated solution and strong/weak aqua are industry terminology. For this guide, the following terminology will be used when discussing solution concentrations.

Weak Absorbent: When referring to both LiBr and aqueous ammonia systems, "weak absorbent" is used in the text. For LiBr systems, "weak absorbent (dilute solution)" is used in the text. For aqueous ammonia systems, "weak absorbent (strong aqua)" is used in the text.

Strong Absorbent: When referring to both types of systems, "strong absorbent" is used in the text. For LiBr systems, "strong absorbent (concentrated solution)" is used in the text. For aqueous ammonia system, "strong absorbent (weak aqua)" is used in the text.

This naming convention allows discussion with industry representatives using their terminology, while maintaining consistent ASHRAE conventions to show similarities of all absorption systems. Although in some cases this becomes cumbersome, it was chosen as the best method to satisfy all practioners. It is similar to dual IP and SI units.

5.3.3 Other Solutions

Other refrigerant-absorbent mixtures exist and are being developed. However, these are currently not commonly used in commercially available absorption systems. These mixtures are briefly discussed in chapter 11.

5.4 CHAPTER LISTING OF DIFFERENT SYSTEMS

Since LiBr and aqueous ammonia systems each have unique requirements, the discussion of these systems is divided into separate chapters. By devoting separate chapters to each system, the unique features of each system can be discussed in detail. Once an absorption approach is selected, only the information pertaining to that system need be referenced.

5.4.1 Lithium Bromide-Water System

A LiBr-water system is typically the best match for an application with cooling applications down to 40°F (6°C). The following chapters are useful for more detailed information on LiBr systems.

Chapter 6: *Lithium Bromide-Water Absorption Cycle*
Chapter 8: *Matching Heat Sources, Cooling Load, and Heat Rejection Capability with an Absorption Machine*
Chapter 9: *LiBr Absorption Machine Applications*

5.4.2 Aqueous Ammonia System

An aqueous ammonia system is typically the best match for an application with refrigeration applications down to –60°F (–51°C). The following chapters are useful for more detailed information on ammonia systems.

Chapter 7: *Aqueous Ammonia Absorption Cycle*
Chapter 8: *Matching Heat Sources, Cooling Load, and Heat Rejection Capability with an Absorption Machine*
Chapter 10: *AAR Machine Applications*

REFERENCES

ASHRAE. 1988. *1988 ASHRAE handbook—Equipment*, chapter 13, p. 13.1. Atlanta: American Society of Heating, Refrigerating and Air-Conditioning Engineers, Inc.

ASHRAE. 1994. *1994 ASHRAE handbook—Refrigeration*, chapter 40. Atlanta: American Society of Heating, Refrigerating and Air-Conditioning Engineers, Inc.

CHAPTER 6
Lithium Bromide-Water Absorption Cycle

In this chapter, the basics of the lithium bromide-water (LiBr) absorption cycle are introduced. Topics include the pressure-temperature-concentration (PTX) equilibrium chart, solution cycles, individual components, chemical additives, part-load operation, and maintenance. This chapter forms the foundation for understanding and applying a LiBr system. The specific requirements for applying a LiBr absorption system are found in chapter 9. Several important features of LiBr machines include:

- LiBr is the absorbent and water is the refrigerant.
- LiBr is a salt that can cause corrosion. Chemical additives reduce or eliminate corrosion potential.
- Crystallization, a problem in the past, has substantially been reduced through the use of microprocessor and direct digital controls (DDC).
- The system operates at a high vacuum.
- Purge systems maintain the vacuum integrity of the machine and remove noncondensables from the machine.

LiBr absorption systems are categorized by the number of times the solution is heated to produce refrigerant vapors. This is referred to as the number of effects. Most LiBr absorption systems are single effect or double effect. A single-effect system uses the input heat once, and a double-effect system uses the heat input for one deabsorption effect and uses the warm refrigerant vapors as the heat source for the second effect.

6.1   PTX EQUILIBRIUM CHART

Since a LiBr absorption machine uses a refrigerant-absorbent solution instead of mechanical compression to transform the low-pressure refrigerant vapor to a higher pressure refrigerant vapor, a pressure-enthalpy (P-H) diagram cannot be used to graphically show the absorption cycle. Instead, a chart that includes the solution concentration—a PTX equilibrium chart—has been developed.

A generic PTX equilibrium chart is shown in Figure 6-1. Highlighted are the relationships among the refrigerant vapor pressure, solution temperature, refrigerant temperature, and the concentration of the solution. An actual LiBr equilibrium chart is shown in Appendix A.

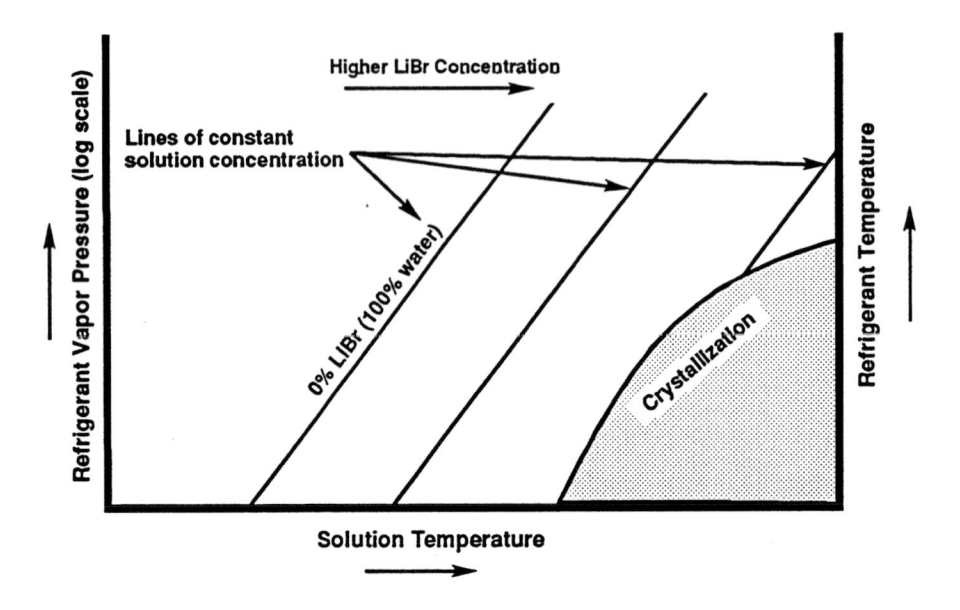

Figure 6-1 Generic PTX Chart

6.1.1 Solution Temperature

The solution temperature is the actual temperature of the LiBr-water solution. In the PTX equilibrium chart, the solution temperature is the x-axis. As the solution temperature increases at a constant pressure and refrigerant temperature, the solution concentration also increases.

6.1.2 Solution Concentration

The solution concentration, represented by the lines sloping up and to the right on a PTX equilibrium chart, indicates the amount of LiBr in the solution. As the refrigerant (water) is separated from the solution, the concentration of LiBr increases (strong absorbent [concentrated solution]). As the refrigerant is absorbed into the solution, the concentration of LiBr decreases (weak absorbent [dilute solution]).

6.1.3 Refrigerant Temperature

The refrigerant temperature corresponds to the temperature of pure refrigerant at the given refrigerant vapor pressure. In the PTX equilibrium chart, the refrigerant saturation temperature is shown on the y-axis.

6.1.4 Refrigerant Vapor Pressure

The refrigerant vapor pressure is also shown on the y-axis using a log scale. The vapor pressure of a refrigerant is defined as the pressure of a pure substance where the liquid and vapor phases coexist in equilibrium. Desorption of the refrigerant occurs when the vapor pressure is increased to an equivalent temperature sufficiently higher than the available cooling water temperature to condense the refrigerant in the condenser. In the absorption cycle, this is the pressure of the generator. The vapor pressure can be increased by raising the temperature of the liquid. With air removed from the absorption system, the vapor pressure of the refrigerant indicates the saturation temperature of the refrigerant.

6.2 SINGLE-EFFECT SYSTEM

This section describes the compressor function of a LiBr absorption machine in the absorption cycle. The condenser, expansion device, and evaporator components of the LiBr absorption machine were discussed in chapter 5 and perform the same functions as those used in vapor-compression systems. The expansion-metering device is much simpler, since the pressure differences are small. The metering can be achieved with an orifice (hole) and the height of liquid refrigerant. Condensers and evaporators in absorption systems typically require pumping and spray headers to ensure wetted surfaces and heat transfer. A schematic of the single-effect absorption machine is shown in Figure 6-2.

6.2.1 Single-Effect Absorption Process

For the refrigeration cycle to work, the refrigerant vapor must change pressure between points 1 and 2. In an absorption machine, the thermal compressor achieves this pressure change.

The absorption cycle is shown on the PTX equilibrium chart in Figure 6-3. The numbers on the PTX equilibrium chart correspond to those in Figure 6-2. At point 1, the refrigerant vapor leaves the evaporator and enters the thermal compressor. At point 2, the refrigerant vapor leaves the thermal compressor and enters the condenser.

Within the thermal compressor, the refrigerant vapor is first absorbed into the LiBr-water solution in the absorber section. Then the now-weak absorbent (dilute solution) is pressurized by the solution pump to overcome the pressure of the generator. In the generator, the refrigerant is separated from the LiBr-water solution by evaporation (heating). This produces a pure refrigerant vapor and a strong absorbent (concentrated solution). A detailed description of the absorption cycle follows.

Path 1→20: At point 1, low-pressure, low-temperature refrigerant vapor from the evaporator enters the thermal compressor into the absorber section.

Path 20→21: From point 20 to 21, the refrigerant vapor is absorbed by the strong absorbent (concentrated solution) from the generator section. Absorption of the refrigerant vapor increases the amount of refrigerant in the solution, thus diluting the solution at point 21 to a weak absorbent (dilute solution).

As the refrigerant vapor is absorbed, it condenses, releasing heat related to the change in phase. In addition, heat is generated by the process of absorption (heat of reaction).

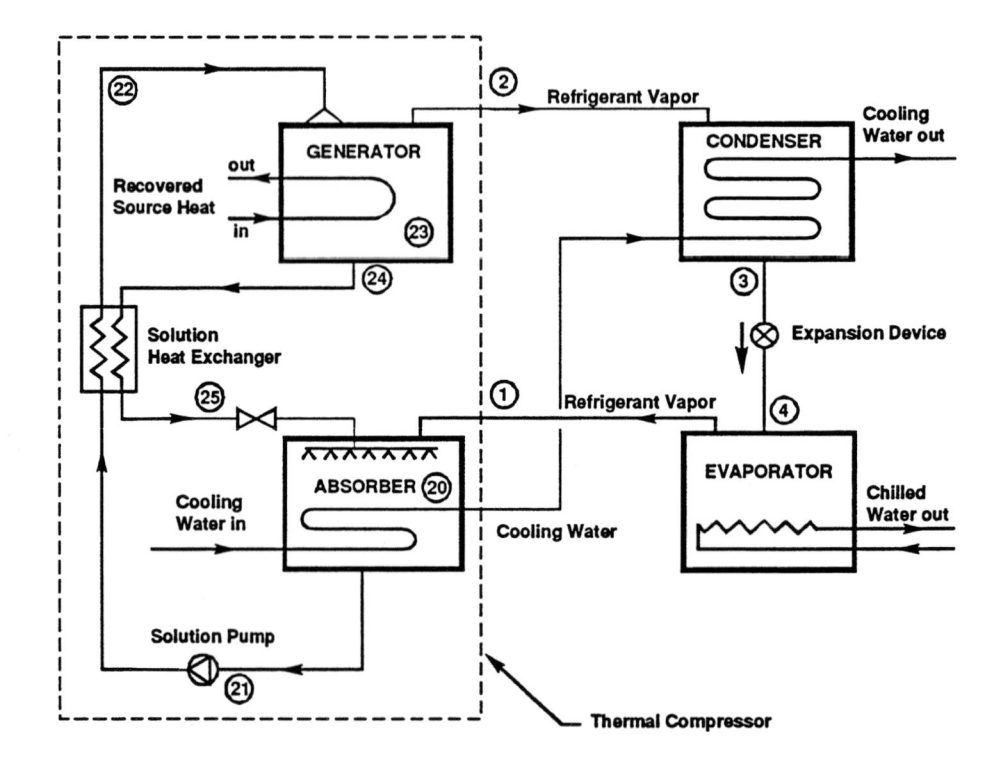

Figure 6-2 Absorption Refrigeration Cycle Schematic, Single-Effect

To remove the heat generated by these two processes, cooling water is circulated through the absorber section. This is required to maintain pressure-temperature-concentration control.

Path 21→22: From point 21, the weak absorbent (dilute solution) is pumped from the absorber section to the generator. In addition to transporting the solution, pumping also overcomes the pressure differences between the absorber and generator sections.

LITHIUM BROMIDE-WATER ABSORPTION CYCLE

The weak absorbent (dilute solution) passes through the solution heat exchanger, where it is preheated to point 22. This heat exchange is a key element in reducing heat energy required for the generator and in increasing cycle efficiency.

Path 22→23: At conditions from point 22, the weak absorbent (dilute solution) enters the generator. Heat is added to the generator from the recovered heat stream to heat the solution to the required temperature.

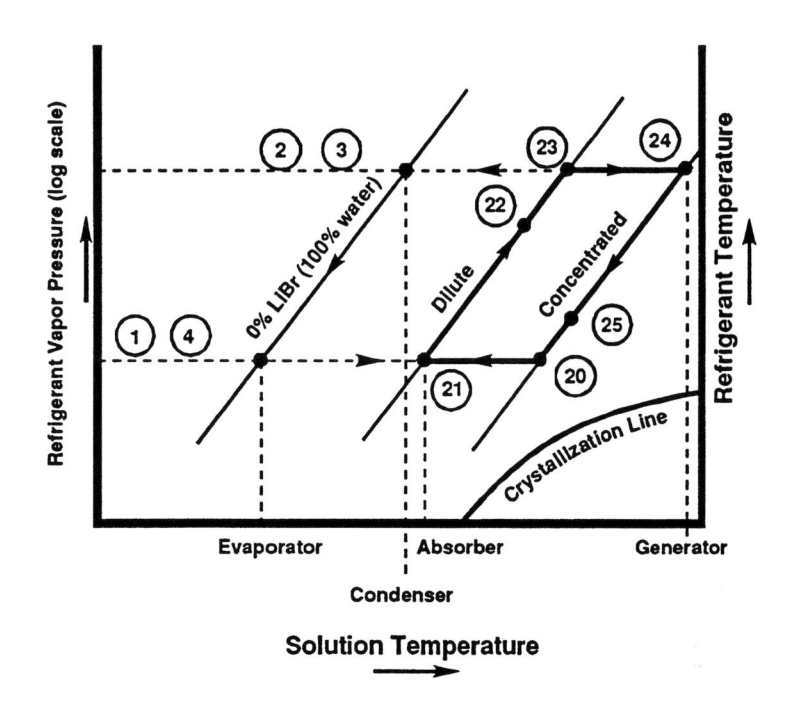

Figure 6-3 PTX Chart, Single-Effect

Path 23→24: When additional heat is added to the solution, the solution boils, releasing refrigerant until the absorbent-refrigerant solution matches the saturated temperature-pressure-concentration related to point 24. At the higher pressure, high-temperature refrigerant leaves the thermal compressor for the condenser at point 2.

With the reduced amount of refrigerant in the solution, the solution becomes concentrated. This is point 24.

Path 24→25: From the generator, the strong absorbent (concentrated solution) passes through a solution heat exchanger, where it is precooled to point 25. The importance of the solution heat exchanger is discussed later in this chapter.

Path 25→20: From point 25, the strong absorbent (concentrated solution) passes through a throttling or expansion device to reduce the pressure of the strong absorbent (concentrated solution). Some of the refrigerant within the solution flashes to vapor.

Water can be economically used as a refrigerant in an absorption machine due to the machine operating at pressures well below atmospheric conditions. In the low-pressure environment of the evaporator in an absorption machine—0.012 psia (82.7 Pa)—water boils at 40°F (4°C) as compared to 212°F (100°C) at atmospheric conditions—14.7 psia (101.3 kPa).

6.2.2 Single-Effect Absorption Machine Configuration

The actual physical configuration of a LiBr single-effect absorption machine varies with each manufacturer. Currently, there are two popular configurations for a single-effect absorption machine. The first configuration, shown in Figure 6-4, houses the entire process within one shell. The second configuration, shown in Figure 6-5, has a separate shell for the condenser and generator sections of the absorption machine.

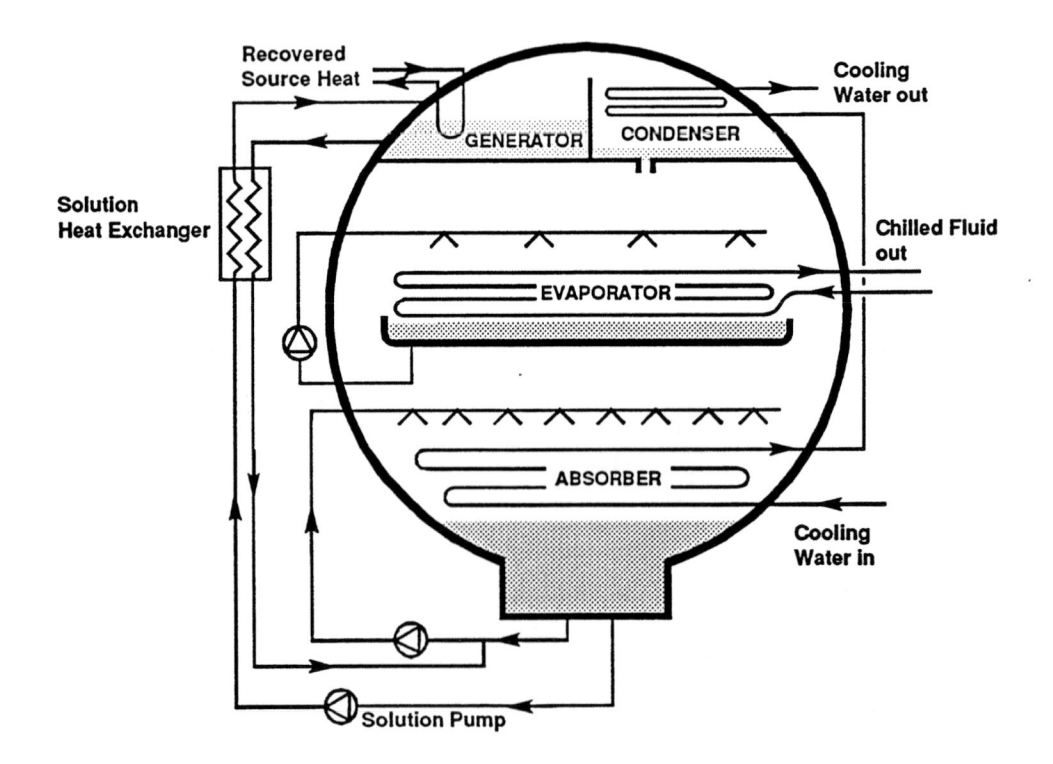

Figure 6-4 Absorption System Layout, Single-Effect, One Shell

LITHIUM BROMIDE-WATER ABSORPTION CYCLE

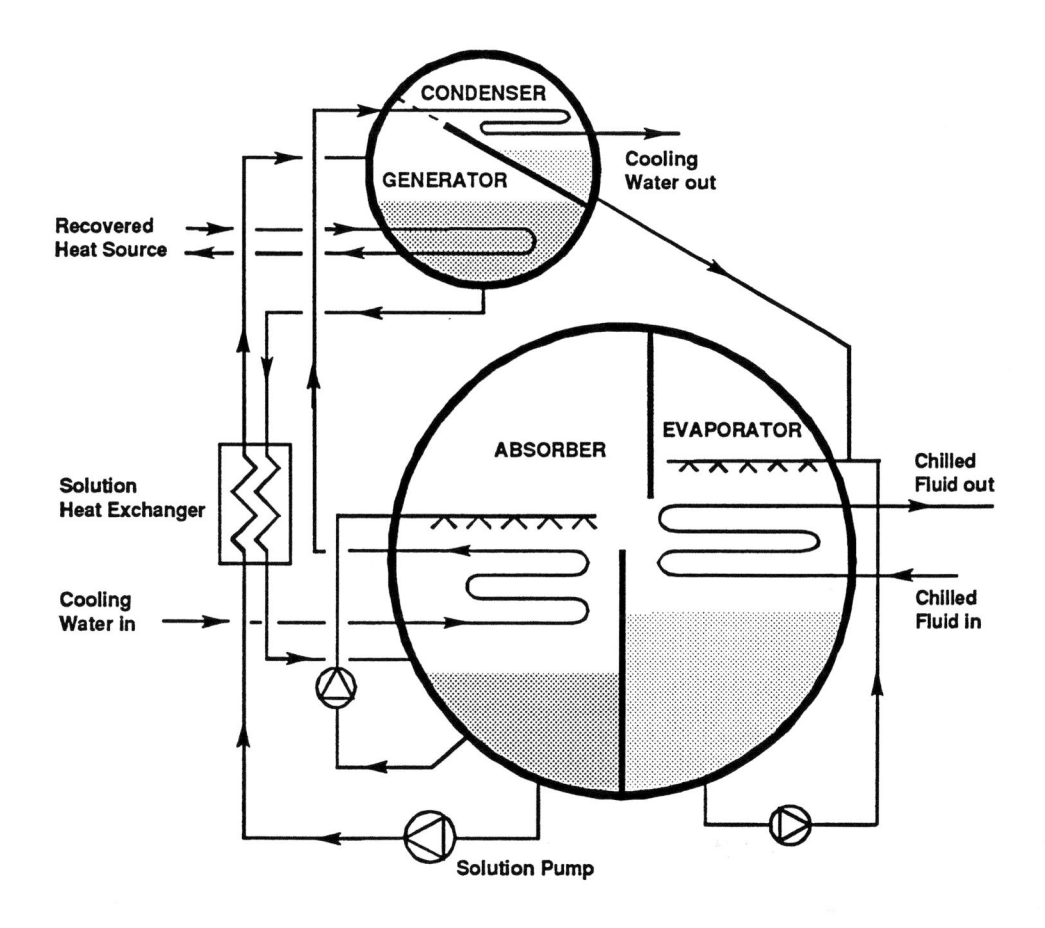

Figure 6-5 Absorption System Layout, Single-Effect, Two Shells

The differences between machines are primarily due to manufacturer preferences, production techniques, and the insulating requirements between separate sections of the absorption machine. The separate sections in an absorption machine are insulated from each other to reduce heat losses and gains between the different sections.

6.3 DOUBLE-EFFECT SYSTEM

A double-effect absorption machine has two stages of generation to separate the refrigerant from the absorbent. The temperature of the heat source required to drive the high-stage generator must be higher than that used for a single-effect machine.

6.3.1 Double-Effect Absorption Process

The overall efficiency of the absorption system is increased by indirectly using the inputted heat a second time. The high-stage generator uses heat supplied by an outside process. The heat from the refrigerant vapor off the high-stage generator is then used to evaporate the refrigerant out of solution in the low-stage generator. This heat transfer is accomplished by passing the high-stage refrigerant vapor through a tube bundle in the low-stage generator. The heat transfer from the high-stage refrigerant vapor to the low-stage solution causes condensation of the high-stage vapor refrigerant (see Figure 6-6).

Cooling water requirements for a double-effect machine are less than those for a single-effect machine, since the heat of condensation for the refrigerant vapor from the high-stage generator is used to desorb the refrigerant from solution in the low-stage generator. Therefore, double-effect machines reject less heat per unit of cooling output. This was shown in Table 3-3 of chapter 3.

There are several popular ways of circulating the weak absorbent (dilute solution) to the high- and low-stage generators. The differences between the methods are primarily due to manufacturer preferences. One way is to first pump the weak absorbent (dilute solution) to the high-stage generator and concentrate it to an intermediate solution concentration (path 21 to 28). The intermediate absorbent is then supplied to the low-stage generator and is further concentrated to a strong absorbent (concentrated solution), path 28 to 24. This series flow type of system is shown schematically in Figure 6-6 and is shown on a PTX equilibrium chart in Figure 6-7.

Another way is to directly supply the weak absorbent (dilute solution) to both the high-stage (path 21 to 28) and low-stage (path 21 to 24) generators. Within each of the generators, the weak absorbent (dilute solution) is concentrated and returned to the absorber section. This parallel-flow system is shown schematically in Figure 6-8 and is shown on a PTX equilibrium chart in Figure 6-9.

Point 29 approaches point 23 in a series flow configuration, and point 29 mixes with point 24 in a parallel-flow configuration as more efficient systems are used. There are more ways to circulate the solution to the generators. Individual manufacturers should be consulted for specific information.

LITHIUM BROMIDE-WATER ABSORPTION CYCLE

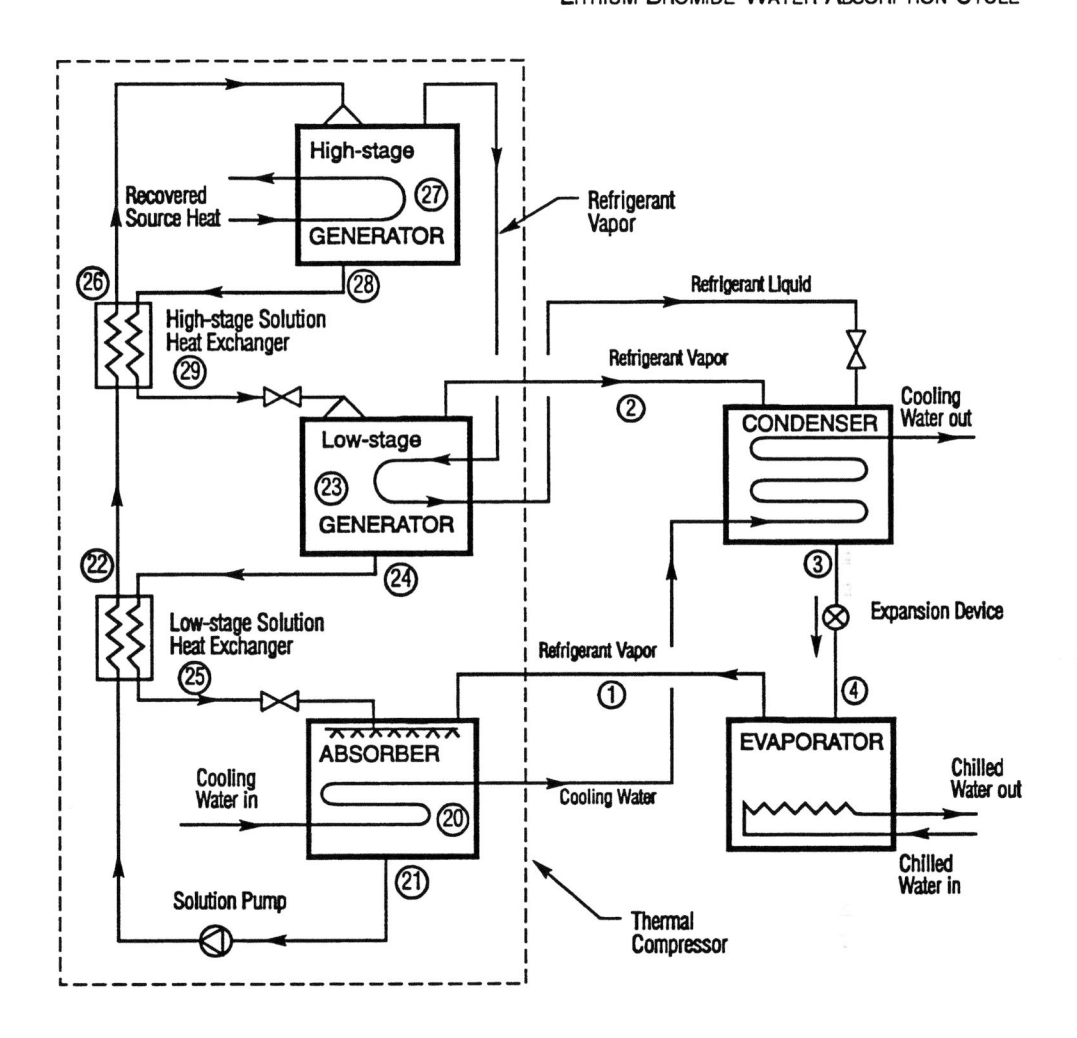

Figure 6-6 Absorption Cycle Schematic, Double-Effect, Series Solution Flow

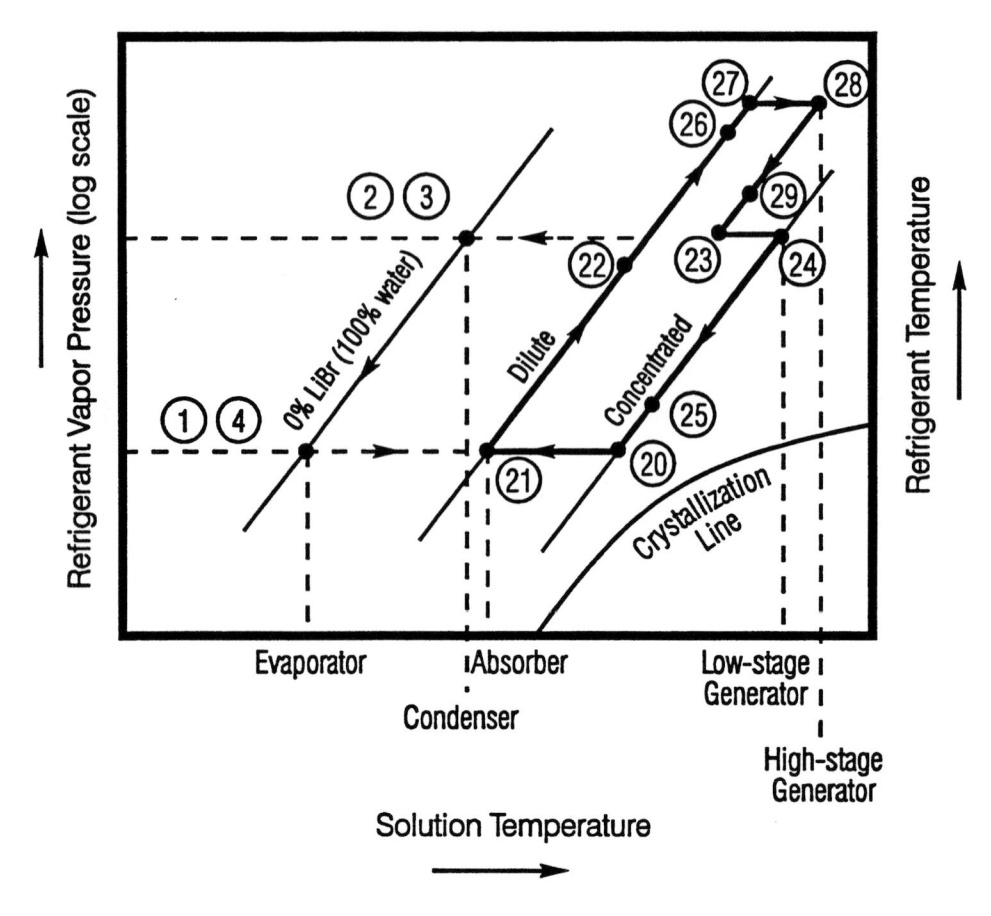

Note:
Heating process 21 to 22 equals cooling from 24 to 24
Heating process 22 to 26 equals cooling from 28 to 29

Figure 6-7 PTX Chart, Double-Effect, Series Solution Flow

LITHIUM BROMIDE-WATER ABSORPTION CYCLE

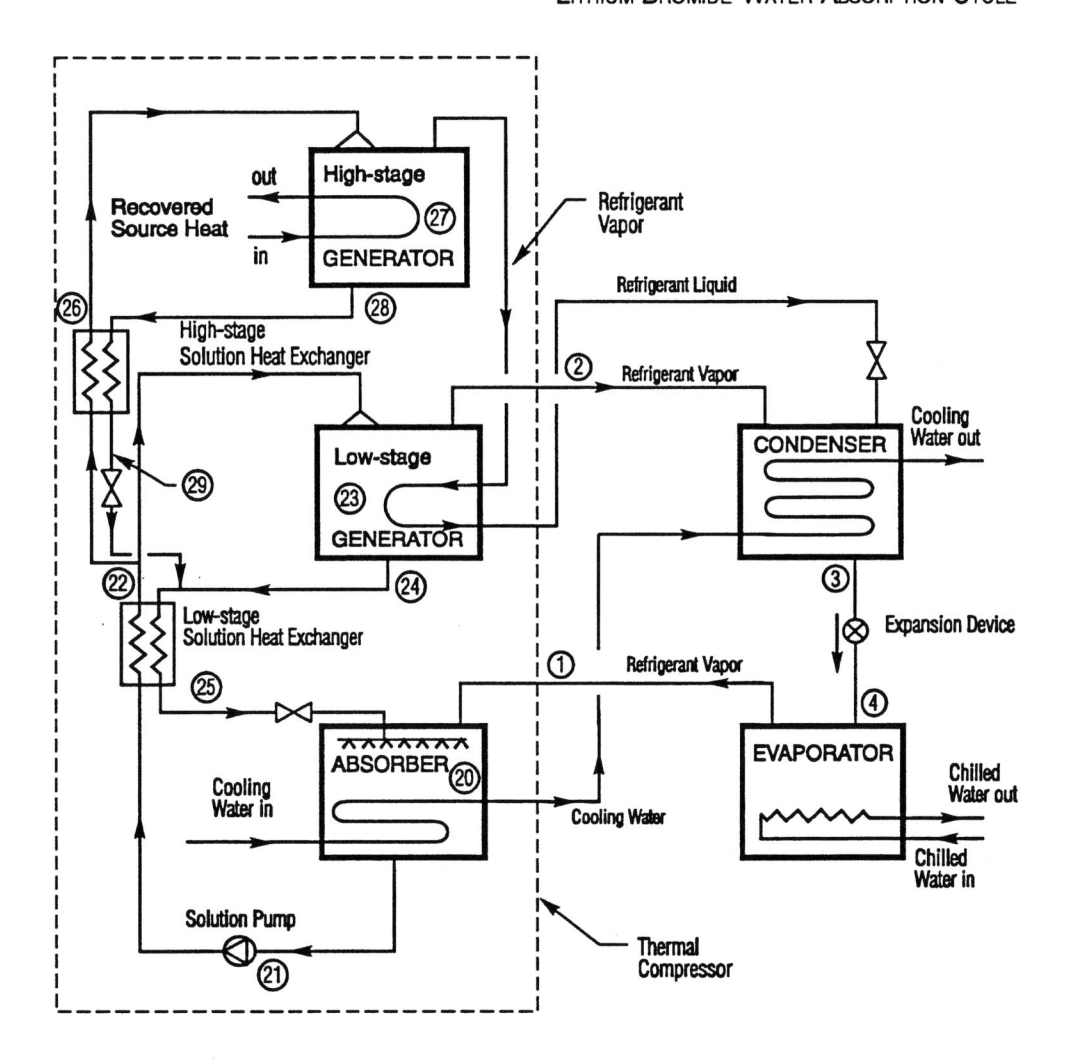

Figure 6-8 Absorption Cycle Schematic, Double-Effect, Parallel Solution Flow

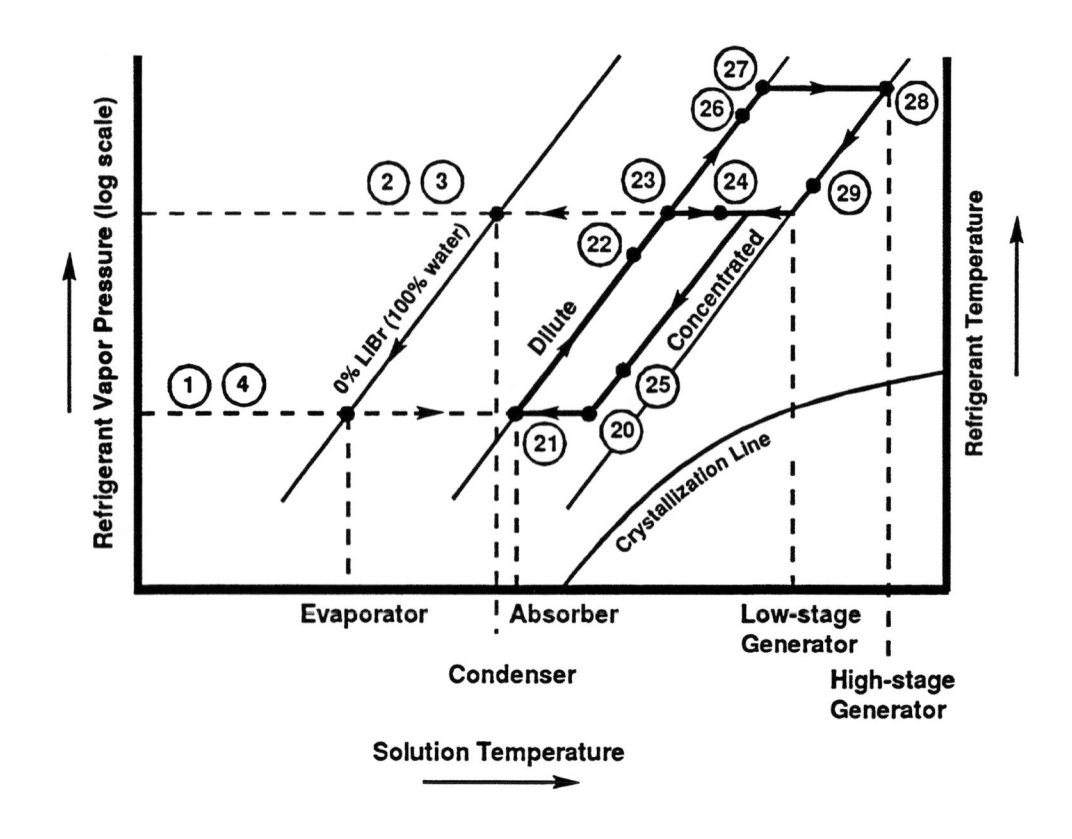

Figure 6-9 PTX Chart, Double-Effect, Parallel Solution Flow

6.3.2 Double-Effect Machine Configuration

As with a single-effect system, the actual physical configuration of the double-effect system varies with each manufacturer. Two of the more popular forms are a double shell and a triple shell. The double-shell variety (Figure 6-10) has one shell for the absorber, condenser, evaporator, and low-stage generator and another shell for the high-stage generator. The triple-shell variety (Figure 6-11) has a separate shell for the condenser and low-stage generator. Within Figures 6-10 and 6-11, the parallel-flow configuration is shown.

6.4 COMPARISON OF SINGLE- AND DOUBLE-EFFECT SYSTEMS

Generally, a double-effect system is chosen over a single-effect system whenever there is a heat source with the temperatures sufficient to power a double-effect machine. Even though the first cost for a double-effect machine is greater than that for a single-effect

LITHIUM BROMIDE-WATER ABSORPTION CYCLE

machine, when recovered heat is utilized, the COP and cooling output are typically maximized with a double-effect machine.

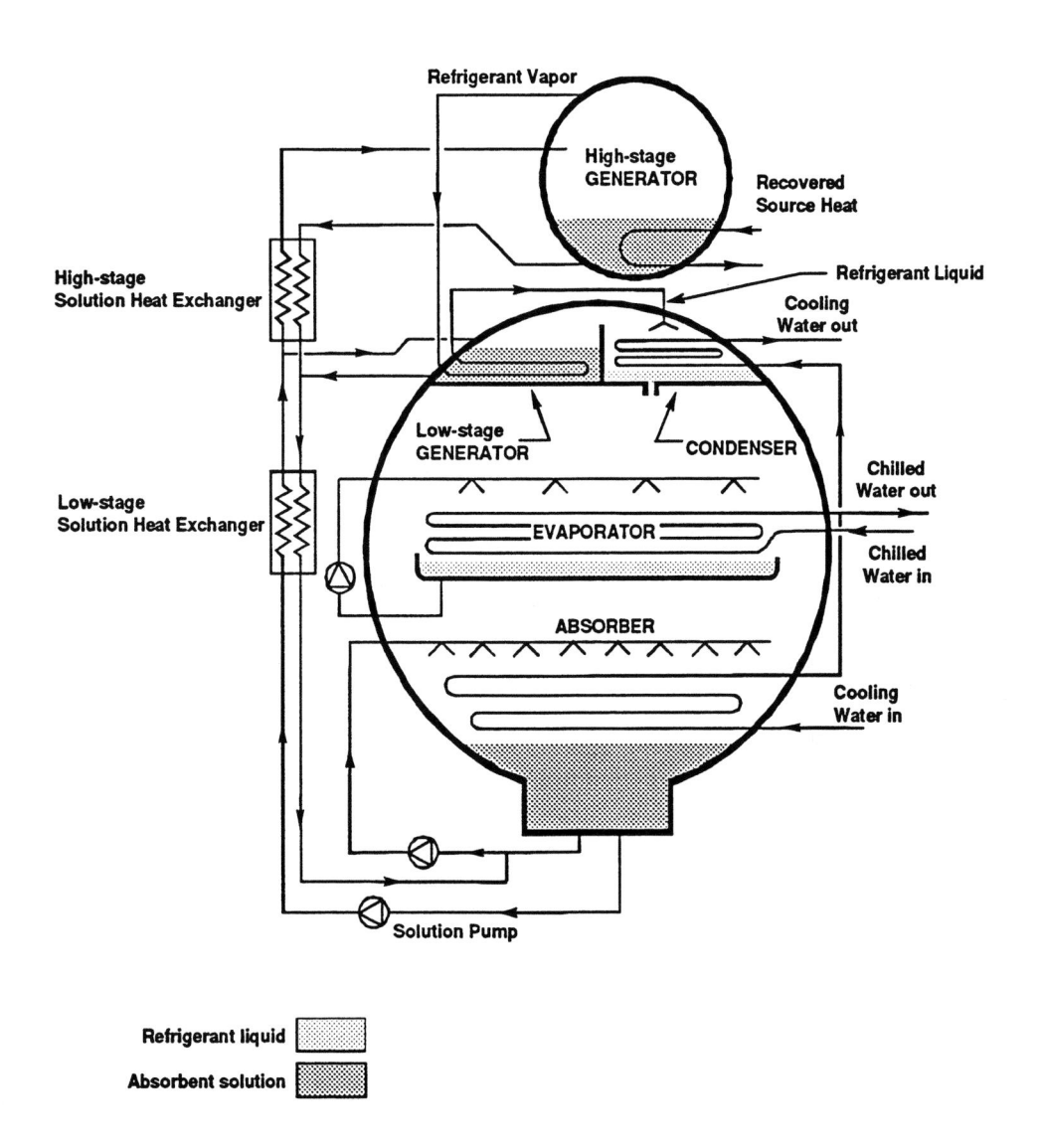

Figure 6-10 Absorption System Layout, Double-Effect, Two Shells, Parallel Solution Flow

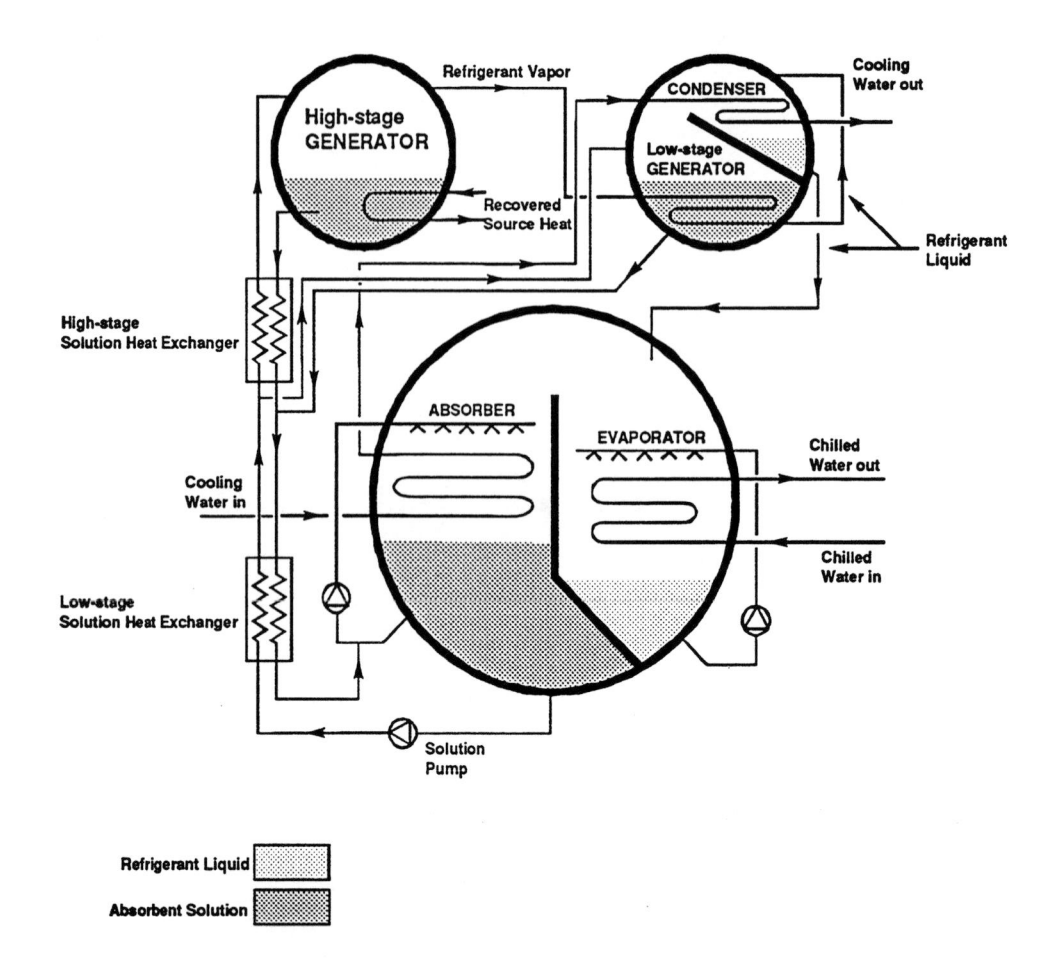

Figure 6-11 Absorption System Layout, Double-Effect, Three Shells, Parallel Solution Flow

A summary of the requirements and parameters of single- and double-effect machines is presented in Table 6-1.

TABLE 6-1
Comparison of Single- and Double-Effect Machines

	Single-Effect LiBr Machine	Double-Effect LiBr Machine
Steam input, psig (kPa)	5–15 (135–205)	65–145 (110–120)
Steam input, °F (°C)	225–250 (110–120)	305–365 (175–185)
Hot water input, °F (°C)	240–300 (115–150)	310–400 (155–205)
Cooling output, °F(°C)	40–100 (4–38)	40–80 (4–27)
COP	0.6–0.7	0.9–1.2
Cooling water, gpm/ton (L/s·kW$_T$)	2.45–6.0 (0.04–0.1)	1.8–4.5 (0.03–0.08)
Machine cost, $/ton ($/kW$_T$)	See Note	See Note
Electrical demand, kW/ton (kW/kW$_T$)	0.012–0.030 (0.003–0.009)	0.015–0.021 (0.004–0.006)

Note: See Table 3-1 in chapter 3 for absorption machine costs.

6.5 INDIVIDUAL COMPONENTS

The previous discussion of LiBr absorption machines was simplified to provide a clear discussion. However, there are additional components in LiBr absorption systems that need to be discussed. These include pumps, additional equipment in the absorber section, solution heat exchanger(s), expansion device(s), additional equipment in the evaporator section, the purge system, the cooling water system, and controls.

6.5.1 Pumps

The pumps in an LiBr absorption machine are hermetically sealed and totally enclosed in the machine to minimize the points where air can leak into the machine. If air is allowed to leak in, the overall efficiency and effectiveness of the absorption cycle is degraded. A further discussion of air leakage is found in section 6.5.6. Pumps are cooled using the absorbent or a side stream of liquid refrigerant.

6.5.2 Absorber Section

To promote the absorption process, the solution is circulated within the absorber section using a pump and a spray distribution system. Spraying the solution increases the surface area for contact between the solution, the refrigerant vapor, and the absorber tube bundle. Figure 6-12 shows details of the absorber section, including the distribution pump and spray system.

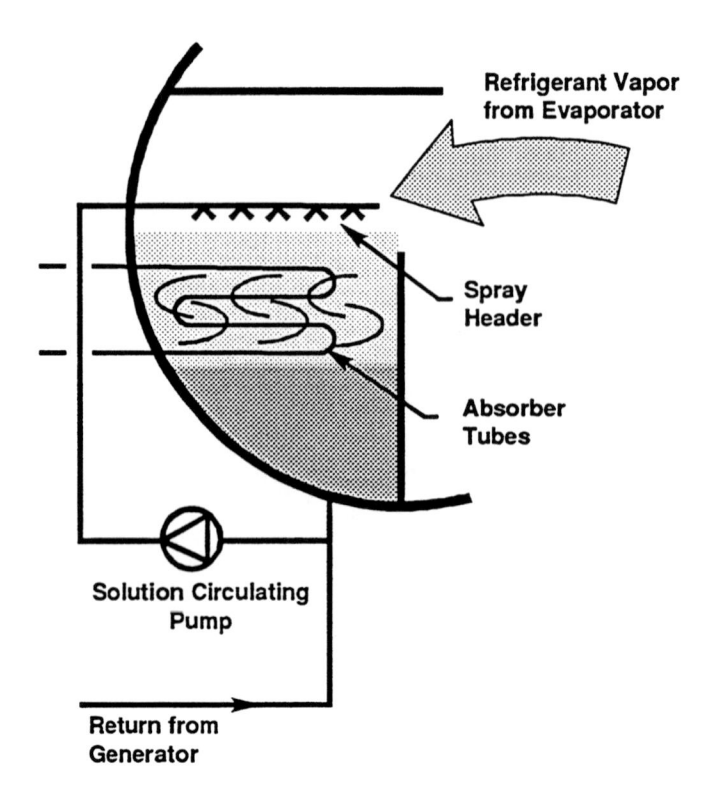

Figure 6-12 Absorber Section Detail

It should be noted that not all absorption machines recirculate the solution in the absorber section. Recirculation can be minimized through the proper design of solution distribution.

6.5.3 Solution Heat Exchanger

The solution heat exchanger precools the strong absorbent (concentrated solution) returning from the generator and preheats the weak absorbent (dilute solution) going to the generator. This reduces the energy required to drive the absorption cycle and the energy that must be rejected from the cycle.

Without the solution heat exchanger, the weak absorbent (dilute solution) would have to be heated in the generator from the absorber temperature to the generator temperature. Also, the strong absorbent (concentrated solution) would have to be cooled by the cooling water system from the generator temperature to the absorber temperature.

The effectiveness of the solution heat exchanger has a large impact on both the cooling capacity and the COP of a machine. The typical effectiveness for solution heat exchangers in LiBr machines is between 70% and 80%. If the effectiveness of the solution heat exchanger decreases from 80% to 60%, the COP is reduced by approximately 0.05 (Manufacturer 1994; Oh et al. 1994).

Figure 6-13 shows the effect the efficiency of the solution heat exchanger has on the energy use of an absorption system.

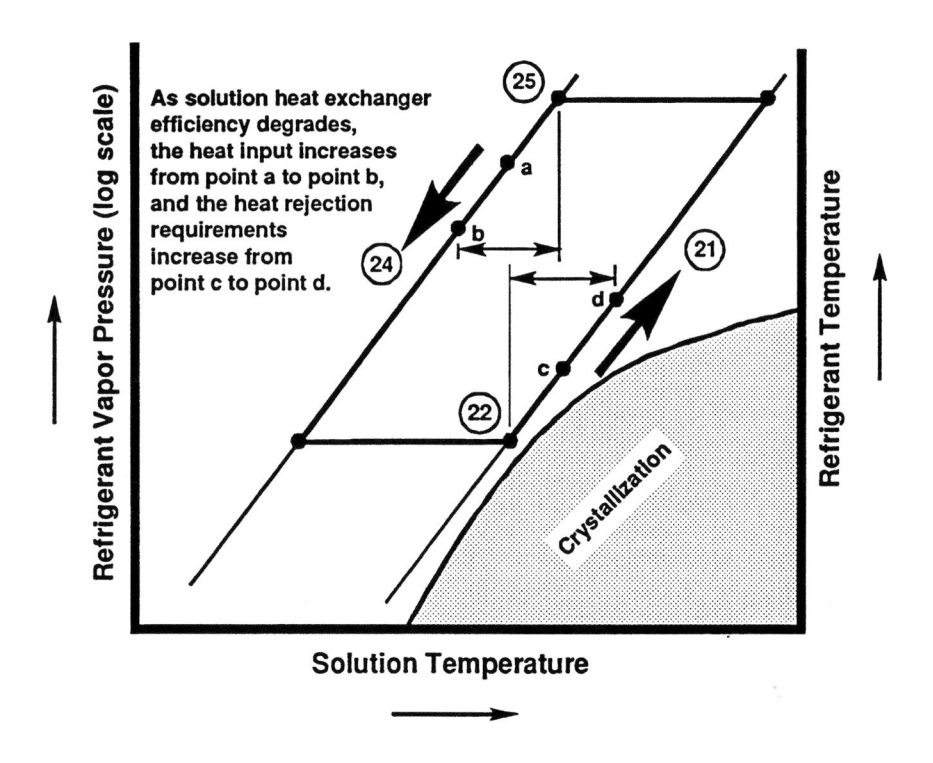

Figure 6-13 Solution Heat Exchanger Degradation Effects

6.5.4 Expansion Device

The purpose of the expansion device is to regulate (meter) the amount of refrigerant flowing from the condenser to the evaporator. The amount of refrigerant circulated determines the cooling effect of the machine.

The expansion process is handled differently by each manufacturer. It is usually an orifice hole or short tubing. The general principle employed is: the more refrigerant liquid in the condenser, the higher the head pressure for driving the liquid through the metering/expansion device, thus the greater the flow of liquid refrigerant through the expansion device.

6.5.5 Evaporator

The evaporator section typically has a circulating pump and a spray distribution system that keeps the liquid refrigerant passing over the evaporator tubes. This ensures a consistent and even heat transfer between the chilled fluid and the refrigerant. This process is shown in Figure 6-14.

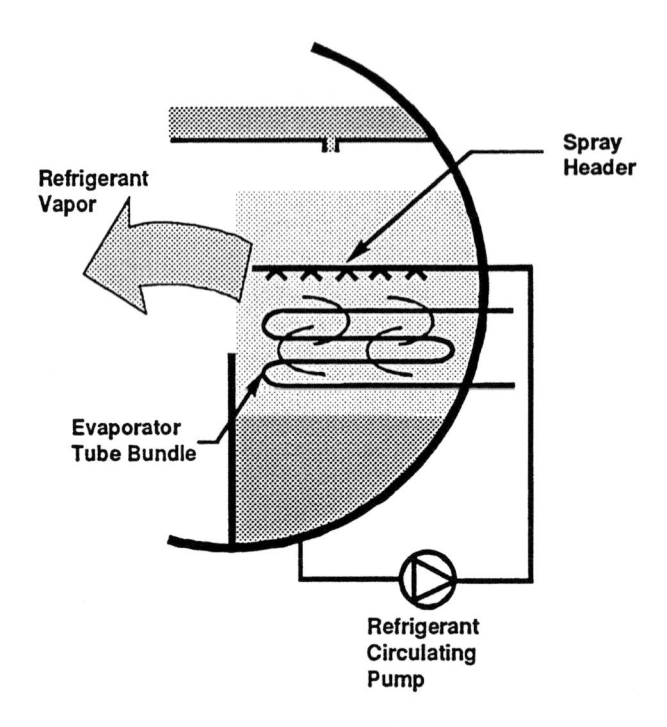

Figure 6-14 Evaporator Section Detail

6.5.6 Purge System

The purge system is operated regularly to remove noncondensable gases (air and hydrogen) from the absorption machine. Air is introduced into the machine through leaks in the machine. Hydrogen is produced from a reaction between the LiBr, water, and steel of the machine. Use of corrosion inhibitors reduces the reaction (Murray 1993).

The noncondensable gases collect in the zones of lowest relative pressure in the absorption machine, which are in the condenser and absorber sections. The majority of noncondensable gases collects in the absorber section. In these locations, the gases form a film between the tubes and the refrigerant (condenser) and between the absorbent and the refrigerant (absorber). The film inhibits the ability of the refrigerant to condense or absorbent to absorb the refrigerant and leads to inefficiency and reduced capacity.

To remove the noncondensable gases, one manufacturer's purge system operates by absorbing a side stream of vapor in a small chamber from the lowest pressure points. The condensable gases (refrigerant) are absorbed out of the side stream and returned to the absorber. The noncondensable gases are pressurized to atmospheric pressure using a vacuum pump and vented. Safety devices ensure air does not enter the absorber system through the purge system.

Another manufacturer has a purge system that operates continuously without the use of a vacuum pump. This type of system uses the solution pump, an educator, and a collection tank to separate the noncondensable gases from the solution. The use of the educator generates a pressure zone that is below that of the absorber or condenser section.

Research has shown that a 5% volume of noncondensable gases reduces the heat and mass transfer rates of the refrigerant vapor in the absorber section by about a factor of 2. To reduce the effect of noncondensable gases to an acceptably low level, approximately 2% to 3%, the concentration of noncondensable gas must be reduced to below 0.1% (Vliet and Cosenza 1991).

The impact of noncondensable gases on the performance of the absorption machine can be significant. For example, if the operating pressure is raised by 0.3 mm Hg due to noncondensable gases, the leaving fluid temperature is raised approximately 1°F (0.5°C). Additional accumulation of noncondensable gases could drive the leaving chilled-water temperature from 44°F (6.6°C) to 50°F (10°C). In an extreme condition, at a cooling part-load of 25%, the system may be operating at full-load conditions since the noncondensable gases raise the leaving chilled-water temperature above the setpoint (Murray 1993).

In addition, noncondensable gases increase the operating cost of an absorption machine by reducing the effective heat and mass transfer areas. Increased energy (heat input) is required to penetrate the film of noncondensable gases (Murray 1993).

6.5.7 Cooling Water System

The cooling water system is critical to the proper operation of an absorption machine. The heat generated in the absorber section and the heat of condensation from the purge section and the condenser section must be removed and rejected to another medium, typically the ambient outdoor air through a cooling tower. The cooling water system typically contains heat rejection equipment, pumps, and associated piping.

6.5.8 Controls

The primary purposes of the control system for LiBr absorption machines are avoiding crystallization, providing capacity control, and ensuring safe and reliable operation of the absorption system.

6.5.8.1 Crystallization

Microprocessor controls, commonly referred to as direct digital controls (DDC), have substantially reduced problems with crystallization. The majority of LiBr systems currently manufactured come with DDC controls that can quickly process large volumes of data and execute corrective action before any problems occur.

In the past, the major complaint with LiBr absorption machines has been crystallization of the system. Crystallization occurs when the LiBr solution freezes, thus blocking passageways and stopping the regenerative process. This stops the cooling effect. The freezing of the solution is a complex reaction where the pumpable liquid solution is converted into a solid or a slushy mass (ASHRAE 1994). The LiBr solution freezes at defined temperatures and concentration combinations.

Figure 6-15 shows the crystallization process of the strong absorbent (concentrated solution) in the solution heat exchanger. With DDC systems, the control system takes corrective action whenever the sensors indicate a possibility of the absorption cycle generating conditions that would allow crystallization to occur. DDC also maintains the cycle efficiency by continuously monitoring system conditions to avoid crystallization. Also included in the controls package are fail-safes, such as bypass valves, to ensure proper system shutdown in the event of power outages or component failures.

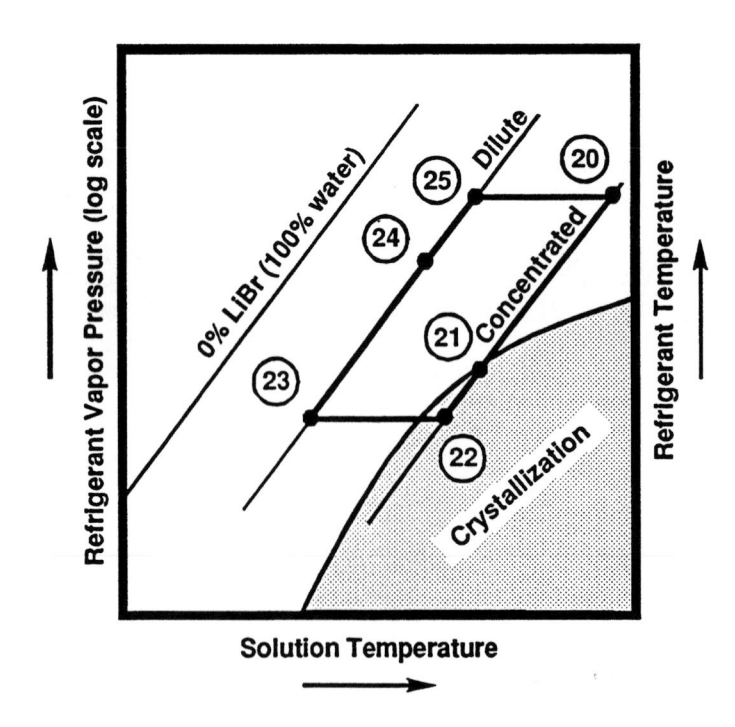

Figure 6-15 PTX Chart with Crystallization Line

LITHIUM BROMIDE-WATER ABSORPTION CYCLE

6.5.8.2 Capacity Control

As the demand for cooling changes, an absorption machine must be able to control its capacity to match the load. Figure 6-16 shows the basic controls required to economically control the capacity of an absorption machine.

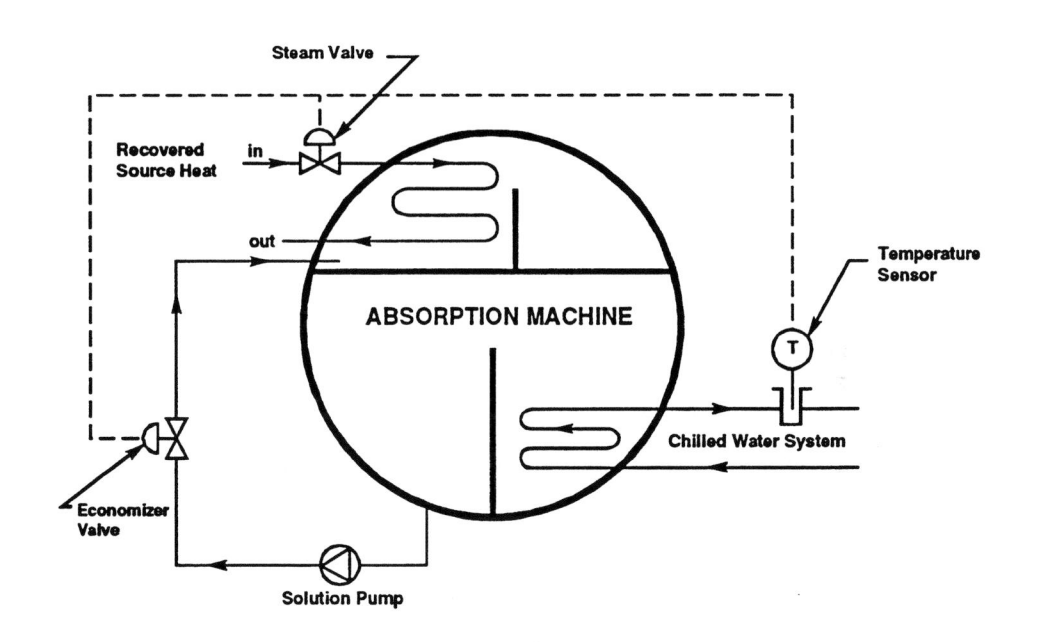

Figure 6-16 Capacity Controls

With a decreasing cooling load, the leaving temperature of the chilled fluid decreases. In response to the decreasing temperature, the valve at the heat inlet begins to close, reducing the amount of heat in the generator. This results in a reduced amount of refrigerant being boiled out of solution. With less refrigerant being generated, constant outlet temperatures can be maintained.

With part loads, an economizer valve in the absorber-generator line closes partially, reducing the amount of weak absorbent (dilute solution) going to the generator. Without the economizer valve, the generator would be required to heat the entire volume of weak absorbent (dilute solution) to boil out the required refrigerant. By reducing the amount of weak absorbent (dilute solution) delivered to the generator, the heat input can be drastically reduced. Figure 6-17 shows the benefits of an economizer valve at part-load conditions (Manufacturer 1985).

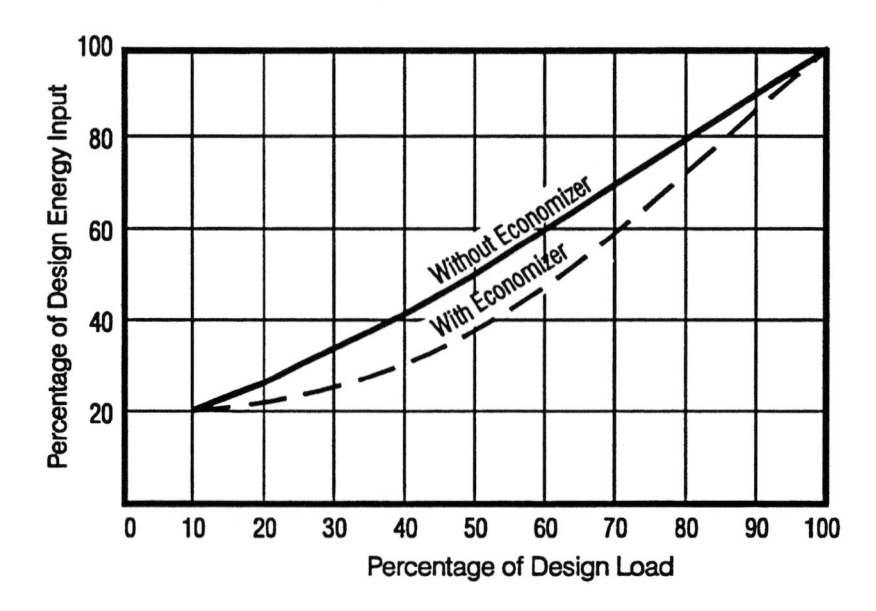

Figure 6-17 Economizer for Part-Load Conditions

6.6 PART-LOAD OPERATION

Since a refrigeration system operates at part load the majority of the time, it is useful to know how this is represented on a PTX equilibrium chart. Figure 6-18 shows a PTX equilibrium chart at full load and at part load for a single-effect machine (Manufacturer 1985, 1994).

As the cooling load on an absorption machine decreases, the amount of heat supplied to the generator decreases and reduces the solution temperature. By decreasing the solution temperature while maintaining the chilled-fluid outlet temperature, the solution becomes more dilute (i.e., more refrigerant stays in the solution and less refrigerant flows through the system). Therefore, part-load conditions move the absorption process shrinks to the left on a PTX equilibrium chart.

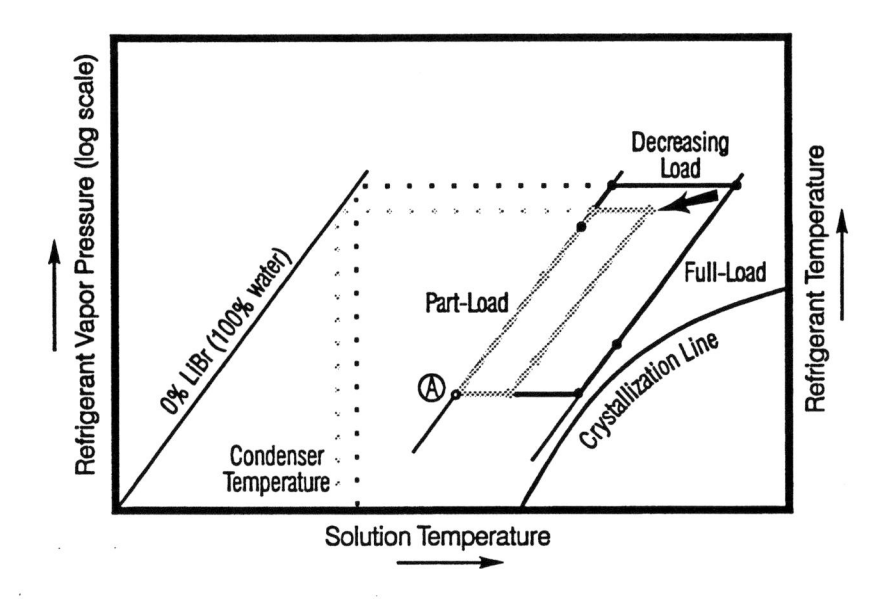

Figure 6-18 PTX Chart, Part-Load Conditions, Single-Effect

In an absorption system, the entering and leaving cooling water temperatures are allowed to float. As the heat input decreases, the condenser heat also decreases. This can result in lower leaving cooling water temperatures, which shrinks the part-load chart vertically. Depending on the control sequences used, the part-load COP can be greater than at full load. It is possible for the entire chart to shift to the left under certain circumstances. However, point A in Figure 6-18 cannot be below the condenser temperature.

The relationship between the percent design load, percent energy input, and the corresponding cooling water temperature for a single-effect system is shown in Figure 6-19. Figure 6-19 shows an interesting correlation in that as the entering cooling water temperature decreases, so does the required heat input to the generator. Therefore, if cooler cooling water is available, the heat input to the absorption machine can be decreased.

Similar to a single-effect machine, under part-load conditions a double-effect machine shrinks to the left on a PTX equilibrium chart. Also, point A cannot move past the condenser temperature line. Figure 6-20 shows a double-effect machine at full load and at part load.

Also, as in a single-effect machine, the energy input is decreased as the load decreases in a double-effect machine. Figure 6-21 shows the relationship between the percent of design load, percent of energy input, and entering cooling water temperature for a double-effect machine.

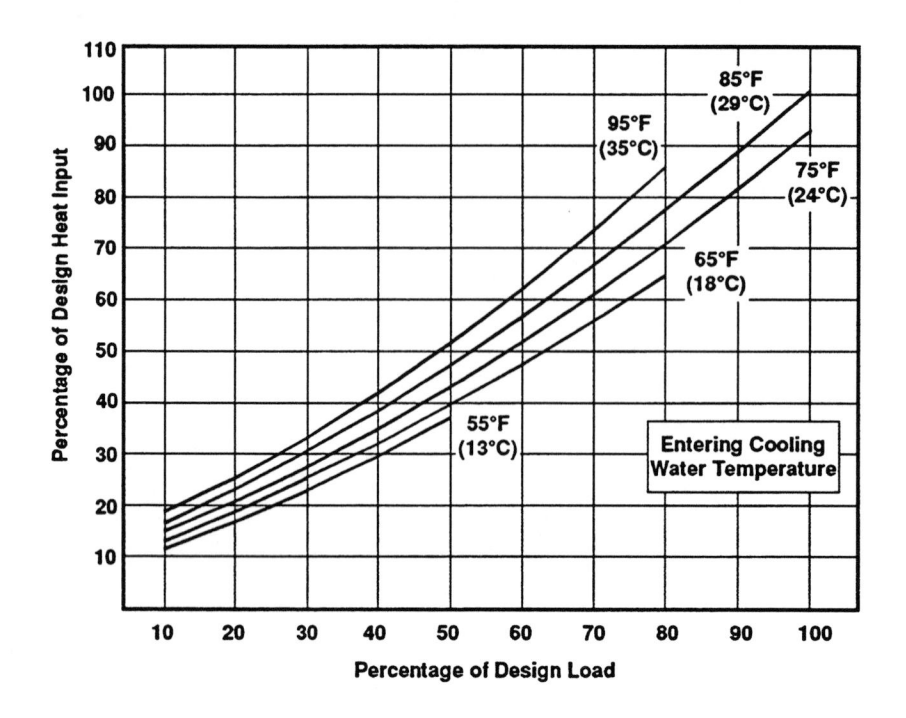

Figure 6-19 Part-Load Capacity Chart, Single-Effect

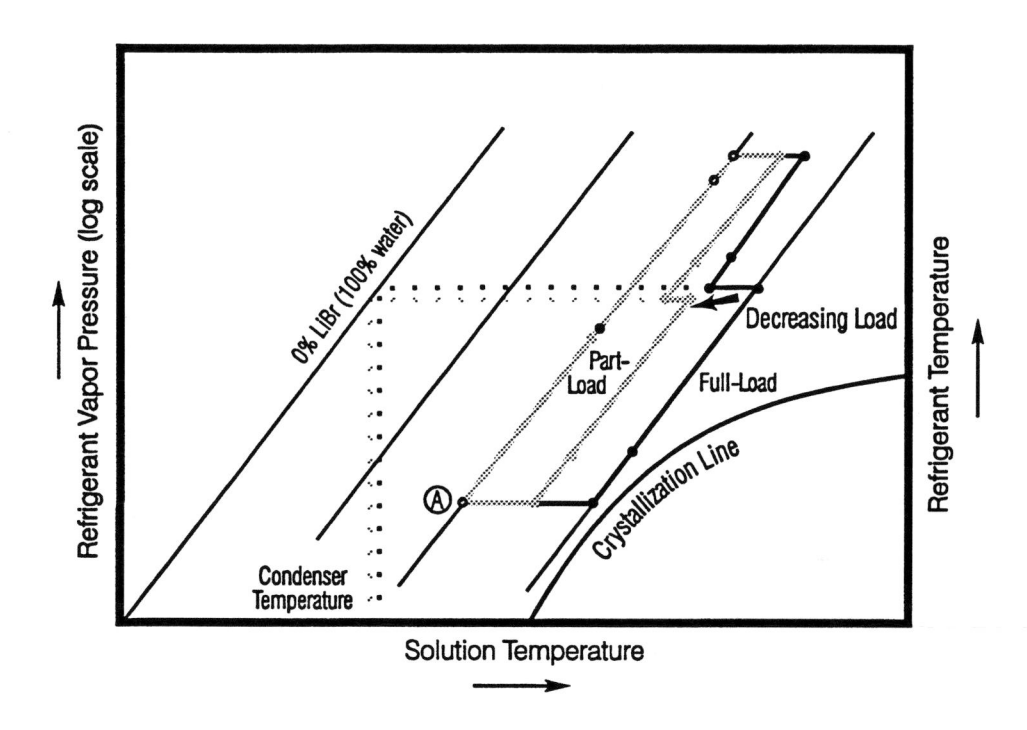

Figure 6-20 PTX Chart, Part-Load Conditions, Double-Effect

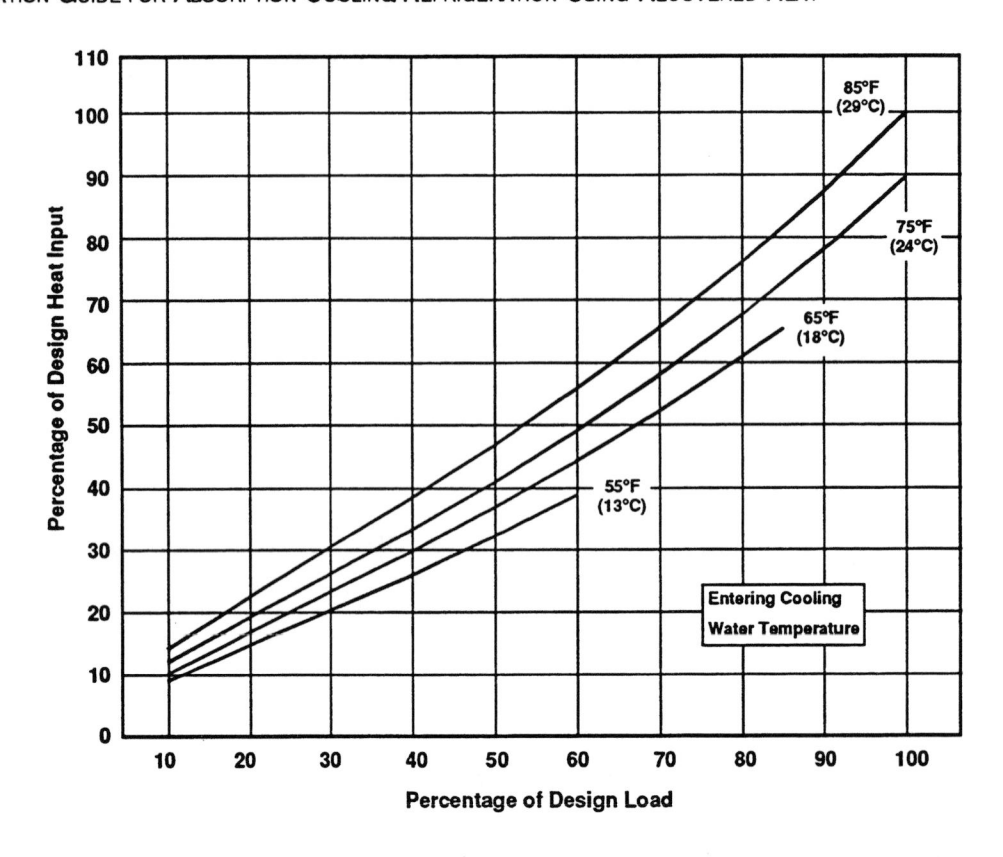

APPLICATION GUIDE FOR ABSORPTION COOLING/REFRIGERATION USING RECOVERED HEAT

Figure 6-21 Part-Load Capacity Chart, Double-Effect

Figures 6-19 and 6-21 are based on a compilation of the available manufacturers' data.

6.7   CHEMICAL ADDITIVES

Chemical additives are used in LiBr absorption machines to reduce the potential for corrosion or to improve the mass/heat transfer of the machine. Additives are also used as pH balancers.

6.7.1   Corrosion Inhibitors

Various manufacturers use different corrosion inhibitors in order to best match the characteristics of the inhibitor with the materials and operating parameters of the absorption machine. The primary corrosion inhibitors in use today include lithium chromate ($LiCrO_4$), lithium molybdate (Li_2MoO_4), and lithium nitrate ($LiNO_3$).

The benefit of a corrosion inhibitor is that the reaction between the lithium bromide and the materials used to construct the machine is greatly reduced. If properly managed, the corrosion inhibitor is safe and effective. However, should the inhibitor be released into the environment, it could have an adverse effect. Inhibitors with chromate are frequently considered a hazardous material. Local regulations will determine concentration and handling requirements for corrosion inhibitors.

Release of the inhibitor would primarily occur when the system is undergoing maintenance, as during normal operation the system is in a vacuum. The inhibitor could also cause a reduction in protection at the higher temperatures used in a double-effect machine. For example, for one of the commonly used inhibitors, an average of 2,400 ppm of the inhibitor is required in the solution to be effective (Reimann 1994). The best defense against corrosion is the proper selection of materials in the manufacture of an absorption machine.

6.7.2 Mass/Heat Transfer Enhancement

As a standard feature of many LiBr absorption machines, 2-ethyl 1-hexanol (C_8H_{18}) is added to the solution to improve the heat and mass (refrigerant) flow of a machine. Typically, the mass flow of the refrigerant is increased by approximately 30% with the addition of C_8H_{18}. This equates to an increase in machine output of approximately 10% to 20%. Typically, 1% of the LiBr solution is composed of C_8H_{18} (Reimann 1994; Jung et al. 1993; Wood et al. 1993).

6.7.3 pH Balancers

The pH of a LiBr absorption machine is balanced to a point using a chemical additive. The balance point and the chemical will vary depending on the machine, the composition of the water used, and the other additives used. Two of the most common chemicals for pH balancing are lithium hydroxide (LiOH) and hydrobromic acid (HBr) (Reimann 1994).

6.7.4 Testing of Solution

Testing of the LiBr solution should be conducted at least once a year for single-effect machines and at least twice a year for double-effect machines. Testing ensures there are no unseen problems with the absorption machine. The testing of the solution is typically accomplished by the factory or factory-approved certified laboratories. Testing can also be accomplished by independent testing laboratories. Whichever testing facility is used, consistency of the results must be ensured from test to test (Reimann 1994). As with any mechanical system, when working with chemicals, follow the handling and use instructions on the material safety data sheet and the instructions from the equipment/chemical manufacturer.

6.8 MAINTENANCE

Regardless of the mechanical system, if proper maintenance is not performed the system degrades. From available information on the maintenance costs for absorption systems, it is estimated that absorption systems cost approximately 0.6 to 1.25 times similarly maintained vapor-compression systems. The differential between the two types of systems has narrowed over the past several years due to the introduction of direct digital control (DDC) for absorption machines and the increased maintenance costs for vapor-compression systems resulting from the CFC recovery requirements.

Due to several factors, such as the corrosiveness of the solution and vacuum pressures, LiBr absorption machines have shorter life expectancies than vapor-compression machines. For properly maintained machines, a LiBr absorption system lasts approximately 70% to 85% as long as a comparable centrifugal machine (Pawelski 1994). Since the quality of machines varies between manufacturers, estimates from the same manufacturer should be used to determine estimated life expectancy of machinery. For example, if a centrifugal vapor-compression machine for a specific manufacturer lasts approximately 30 years, then an absorption machine of equal quality from the same manufacturer should last approximately 20 to 25 years.

REFERENCES

ASHRAE. 1994. *1994 ASHRAE handbook—Refrigeration*, p. 40.2. Atlanta: American Society of Heating, Refrigerating and Air-Conditioning Engineers, Inc.

Dorgan, C. B. 1994. Personal communication with Trane, York, and multiple HVAC contractors nationwide (phone).

Jung, S.-H., C. Sgamboti, and H. Perez-Blanco. 1993. An experimental study of the effect of some additives on falling film absorption. *AES-Vol. 31, International Absorption Heat Pump Conference*, pp. 49–55.

Manufacturer. 1985. *Absorption refrigeration*. Air Conditioning Clinic. No. 2803-11-677. LaCrosse, WI: The Trane Co.

Manufacturer. 1994. Personal communication with the Trane Company (phone).

Murray, J. G. 1993. Purge systems for absorption chillers. *ASHRAE Transactions* 99(1): 1485–1494.

Oh, M. D., S. C. Kim, Y. L. Kim, and Y. I. Kim. 1994. Cycle analysis of air-cooled, double-effect absorption heat pump with parallel flow type. AES-Vol. 31, *International Absorption Heat Pump Conference*, pp. 117–123.

Pawelski, M. 1994. Personal communication (meeting).

Reimann, R. 1994. Personal communication (phone).

Vliet, G. C., and F. B. Cosenza. 1991. Absorption phenomena in water-lithium bromide films. Japanese Absorption Heat Pump Conference, Tokyo, Japan, Conference Proceedings, Sept.–Oct.

Wood, B. D., N. S. Berman, K. Kim, and D.S.C. Chau. 1993. *Heat transfer additives for absorption cooling system fluids*. Prepared for Gas Research Institute, no. 93/0149.

CHAPTER 7
Aqueous Ammonia Absorption Cycle

In this chapter, the basics of an aqueous ammonia absorption refrigeration (AAR) cycle are reviewed. Topics include solution composition, solution cycles, individual components, and chemical additives. This chapter forms the foundation for understanding and applying an AAR system. Specific requirements for applying an AAR system are found in chapter 10.

Several important features of AAR machines include the following:

- Water is the absorbent and ammonia is the refrigerant.
- The cooling output is typically liquid ammonia (separate from the ammonia in the AAR machine) versus the chilled water typically used in LiBr absorption systems. Since ammonia is the refrigerant, the output of an AAR machine can be as low as –60°F (–51°C).
- A fractional distillation process, which separates the ammonia and the water vapor, is used since a small amount of water is converted to the vapor state within the absorption cycle operating temperatures and pressures.
- If installed indoors, ventilation is required by standards for ammonia refrigeration.
- Copper is not used in the construction of AAR systems, as ammonia and copper react adversely with each other.

AAR systems are categorized by the number of evaporator stages. A single-stage system has one stage of evaporation/absorption, and a two-stage system has two stages of evaporation/absorption. The number of stages of evaporation can be increased in order to handle the various cooling and refrigeration temperatures required by the functional process.

AAR machines are custom designed and built for each application. The principles behind the operation of an AAR machine are similar to those of LiBr absorption machines.

7.1 GRAPHICAL REPRESENTATION OF AAR CYCLE

In an AAR absorption cycle, the low-pressure refrigerant vapor from the evaporator is converted to high-pressure refrigerant by using a thermal compressor rather than a mechanical compressor. Unlike vapor compression, which directly pressurizes the refrigerant vapor from low to high pressure, an absorption system uses a secondary substance (absorbent), a pump, and heat to achieve the higher pressure state required for the refrigerant vapors to be condensed. As the properties of the refrigerant/absorbent solution cannot be illustrated directly in a P-H diagram, other charts and tables have been developed to aid in the analysis of the thermal compression cycle.

There are several charts or tables available to determine the properties of the aqueous ammonia solution at the different locations in the AAR machine. There are three common methods used to obtain ammonia-water property data.

Temperature-concentration charts contain information on solution concentration, solution temperature, saturation pressure, enthalpy of saturated vapor, vapor composition, and enthalpy of saturated liquid. A temperature-concentration chart is typically referred to as a Ponchon diagram. See Figure A-2 in Appendix A.

Enthalpy-concentration charts are typically used to describe ammonia-water property data at a single pressure. However, if the high and low pressures are known, the two pressures may be juxtaposed in the same diagram. See Figure A-3 in Appendix A.

Property tables can be developed from the temperature-concentration and enthalpy-concentration charts.

Since the AAR cycle can be explained and applied without describing the solution properties, the graphical format of the AAR cycle is not included in this chapter. Chapter 10 presents a graphical method for optimizing the generator and condenser temperatures of an AAR system. Manufacturers or design-build contractors should be contacted for detailed information on the design of AAR systems.

7.2   ONE-STAGE SYSTEM

A one-stage AAR system has a single stage of evaporation/absorption. A schematic of the one-stage absorption cycle is shown in Figure 7-1. This chapter concentrates on the thermal compressor function of an AAR absorption machine.

For the refrigeration cycle to work, the refrigerant vapor must change pressure to provide the required evaporator and condenser temperature requirements. In the absorption machine, the thermal compressor provides this increase in pressure. The refrigerant vapor leaves the evaporator and enters the thermal compressor at point 1 and leaves the thermal compressor and enters the condenser at point 2.

Within the thermal compressor, pure refrigerant vapor is absorbed into the water-ammonia solution in the absorber section. The solution is pressurized by the solution pump to overcome the pressure of the generator. In the generator and rectifier, pure anhydrous ammonia is separated from the solution by distillation (boiling).

Aqueous Ammonia Absorption Cycle

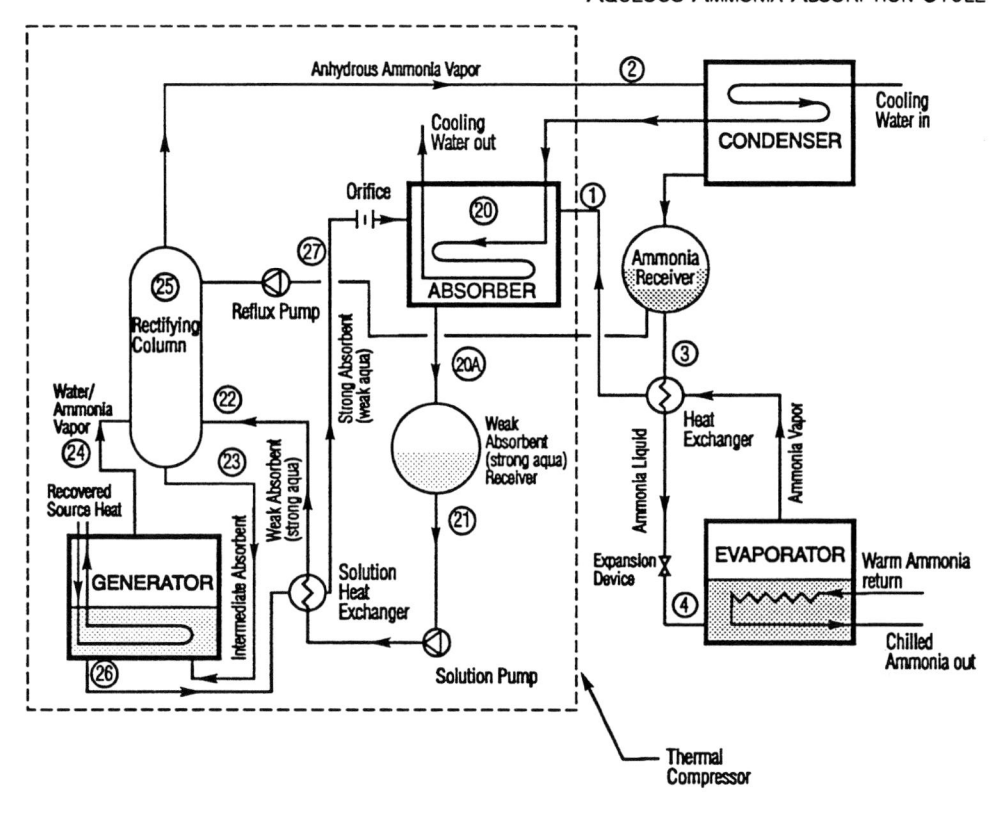

Figure 7-1 Aqueous Ammonia System Schematic, One-Stage

A detailed description of the absorption cycle following the path shown in Figure 7-1 follows.

Path 1→20: At point 1, low-pressure, low-temperature refrigerant vapor from the evaporator enters the thermal compressor into the absorber section.

Path 20→21: From point 20 to 20A, the refrigerant vapor is absorbed by the strong absorbent (weak aqua) from the generator section. Absorption of the refrigerant vapor dilutes the solution and generates the weak absorbent (strong aqua), point 20A.

As refrigerant vapor is absorbed, it condenses, releasing heat. In addition, heat is generated by the process of absorption (heat of reaction). This heat is removed by cooling water circulated through the absorber section. The weak absorbent (strong aqua) drains by gravity into the strong aqua receiver, point 21.

Path 21→22: From point 21, the weak absorbent (strong aqua) is transported from the absorber section to the generator by a pump. In addition to transporting the solution, the pump also overcomes the pressure differences between the absorber and the generator sections. The weak absorbent (strong aqua) passes through the solution heat exchanger and is preheated with warm return solution from the generator to point 22.

Paths 22→23→24→25 and 22→23→26:

The weak absorbent (strong aqua) enters the rectifying column at point 25 and flows to the generator (desorber). At the generator, the weak absorbent (strong aqua) is heated and the ammonia, along with a small amount of water (point 24) is driven out of solution and into the rectifier. In the rectifier, the ammonia is separated from the water by a distillation/rectification process. This process is described in detail in section 7.4.5. The solution exiting the generator is concentrated (weak aqua), point 26.

Pure (anhydrous) ammonia is added to the top of the rectification column (reflux). This accomplishes the final purification process (point 25).

The high-pressure, high-temperature refrigerant leaves the thermal cycle to the condenser at point 2.

Path 26→27: From the generator (point 26), the strong absorbent (weak aqua) passes through a solution heat exchanger, where it is subcooled to point 27. The importance of the solution heat exchanger is discussed later in this chapter.

Path 27→22 From point 27, the strong absorbent (weak aqua) passes through a throttling or expansion device to reduce the pressure of the strong absorbent (weak aqua). The strong absorbent (weak aqua) is subcooled in the solution heat exchanger. Therefore, the solution will not flash to a vapor as it passes through the expansion device.

The actual physical configuration of an AAR machine varies from site to site, as the machines are primarily custom-designed and built for each application. However, the general configuration of the machine is similar to the one shown in Figure 7-1.

7.3 MULTISTAGE SYSTEMS

A typical application for a multistage AAR system is a cold storage warehouse with a cooler and a freezer section. The cooler is typically maintained at 25°F (–4°C) and the freezer at –40°F (–40°C). The most efficient means to provide these two different temperatures to the load is to have separate evaporators at the corresponding approach temperatures. For the cold storage warehouse example, a two-stage AAR system with two stages of evaporation/absorption could be applied.

Figure 7-2 shows a schematic of a two-stage AAR system. Similar to some vapor-compression refrigeration cycles, the high-temperature evaporator feeds the low-temperature evaporator in a cascade fashion. The refrigerant vapor from the low-temperature evaporator (point 7) is supplied to the low-temperature absorber, where it is absorbed into the strong absorbent (weak aqua). The low-temperature weak absorbent (strong aqua) is then pumped

Aqueous Ammonia Absorption Cycle

to the high-temperature absorber as a strong absorbent (weak aqua). The high-temperature absorber performs the same function as a one-stage AAR system absorber.

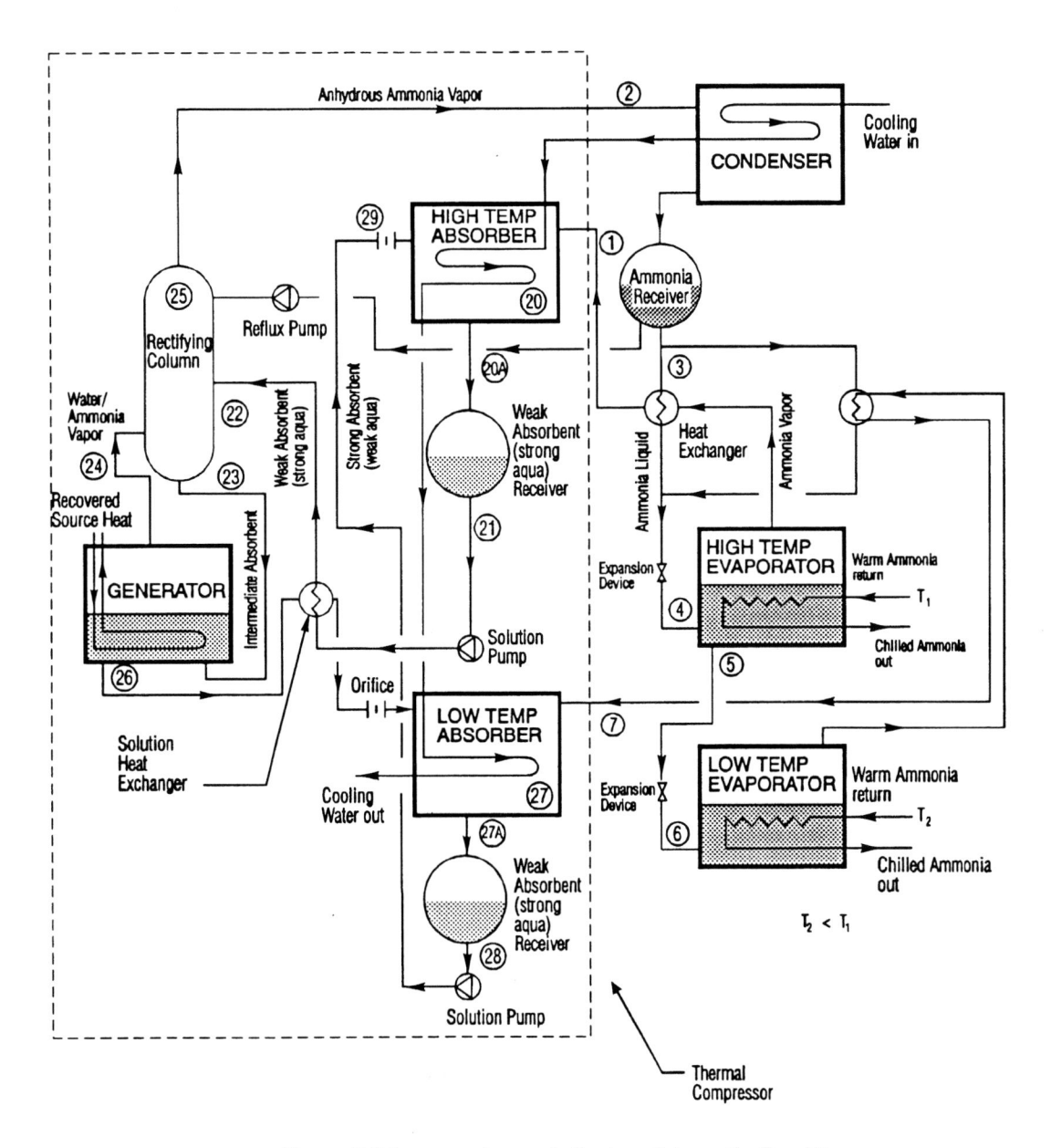

Figure 7-2 Aqueous Ammonia System Schematic, Two-Stage

As with a one-stage AAR system, the physical layout of the two-stage AAR system varies with each location. The general layout will be similar to Figure 7-2. The number of stages of refrigeration can be increased as required by the load variations. The increased efficiency versus the increased first costs must be compared to determine the appropriate number of stages. Maintenance and simplicity of operation are also factors in determining the appropriate number of stages.

7.4 INDIVIDUAL MECHANICAL COMPONENTS

The previous discussion on absorption machines was simplified in order to provide a clear discussion. However, additional components in an absorption system need to be discussed, including the types of pumps used in an absorption system, additional equipment in the absorber and evaporator sections, the solution heat exchanger, the generator/rectifier section, the ammonia condenser, the expansion device, the cooling water system, and controls.

7.4.1 Pumps

The pumps in an AAR machine are typically centrifugal and are used to pressurize and transport the weak absorbent (strong aqua) from the absorber receiver to the generator. Reflux pumps may be required to pressurize and transport pure ammonia liquid from the condenser receiver to the rectifier.

Centrifugal pumps in an AAR system provide for long life and troublefree operation. In a typical system, the absorber/evaporator pressure is approximately 0 psig (100 kPa) and generator/condenser pressures are approximately 195 psig (1,445 kPa) (Shepherd 1994).

7.4.2 Absorber Section

In the absorber section, the refrigerant vapor from the evaporator is absorbed by the strong absorbent (weak aqua). To promote the absorption process, the solution is sprayed from the top of the absorber section. By spraying the solution, the surface area for contact with the solution, the refrigerant vapor, and the absorber cooling tube bundle is increased. Figure 7-3 shows a detail of the absorber section, including the spray system.

AQUEOUS AMMONIA ABSORPTION CYCLE

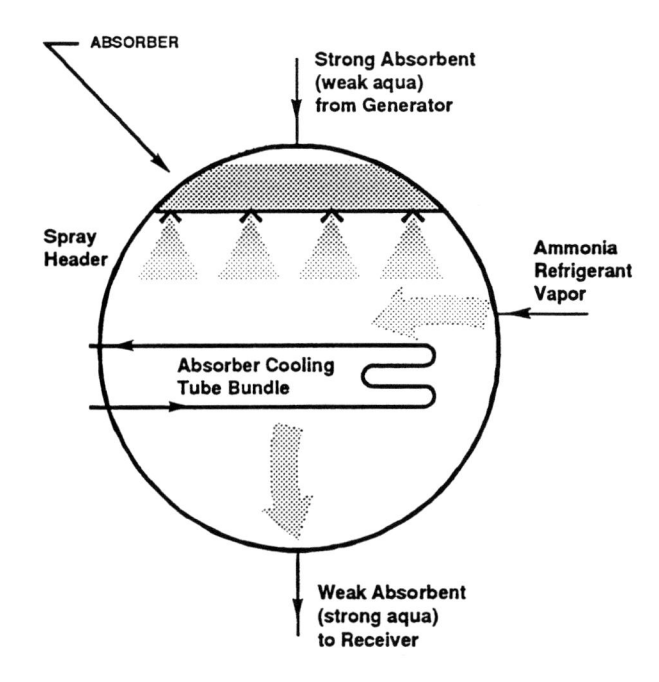

Figure 7-3 Absorber Section Detail

7.4.3 Evaporator Section

Several different types of evaporator systems are employed to ensure proper heat transfer in the evaporator section. Typically, a flooded evaporator is used, with the shell side flooded with the liquid ammonia refrigerant from the AAR system. The refrigerant vapor or liquid from the load side of the system is circulated through the tubes. Figure 7-4 shows details of a flooded evaporator.

The transfer media on the load side of the evaporator can be ammonia, brines, or fluorocarbons. Ammonia is the most popular transfer medium.

7.4.4 Solution Heat Exchanger

The solution heat exchanger precools the strong absorbent (weak aqua) returning from the generator and preheats the weak absorbent (strong aqua) going to the generator. The use of the solution heat exchanger reduces the total heat input required to run the absorption cycle and the amount of heat that must be rejected from the cycle to the cooling water circuit.

Without the solution heat exchanger, the weak absorbent (strong aqua) would have to be heated from the absorber temperature to the generator temperature. Also, the strong absorbent (weak aqua) would have to be cooled by the cooling water system from the generator temperature to the absorber temperature.

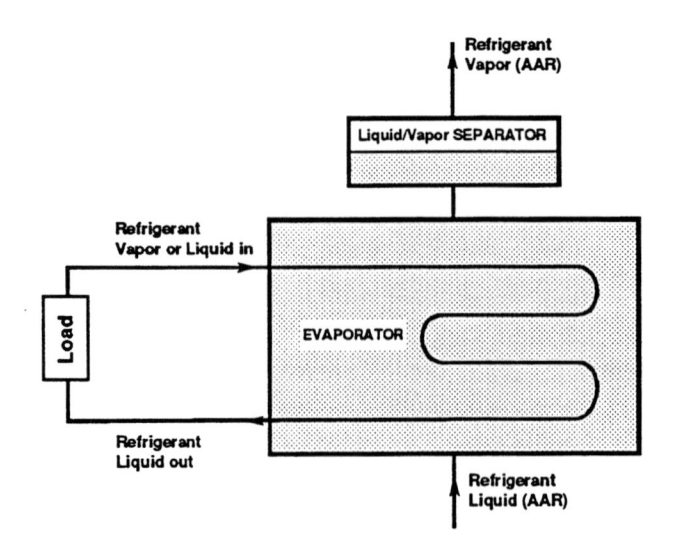

Figure 7-4 Evaporator Section Detail

7.4.5 Generator/Rectifier

In the generator/rectifier portion of an AAR system, refrigerant is separated from the absorbent by heat added in the generator and by distillation in the rectifier. Recovered heat is supplied to the generator, vaporizing the ammonia and some water. This vapor is then directed through a series of distillation trays in which the vapor is fed in a counterflow direction to liquid refrigerant. This process is one of reflux. The liquid refrigerant absorbs the water vapor, resulting in pure (99.96%) ammonia vapor leaving the rectifier. This process is shown in Figure 7-5.

Weak absorbent (strong aqua) from the absorber receiver and the water/ammonia vapor enter the rectifying column at the bottom (points 22 and 24). The liquid ammonia enters the rectifying column at the top (point 25). The multiple trays contain an intermediate-strength absorbent (aqua). The ammonia vapor exits the rectifier at the top (point 25) and the intermediate absorbent exits the rectifier at the bottom (point 23).

AQUEOUS AMMONIA ABSORPTION CYCLE

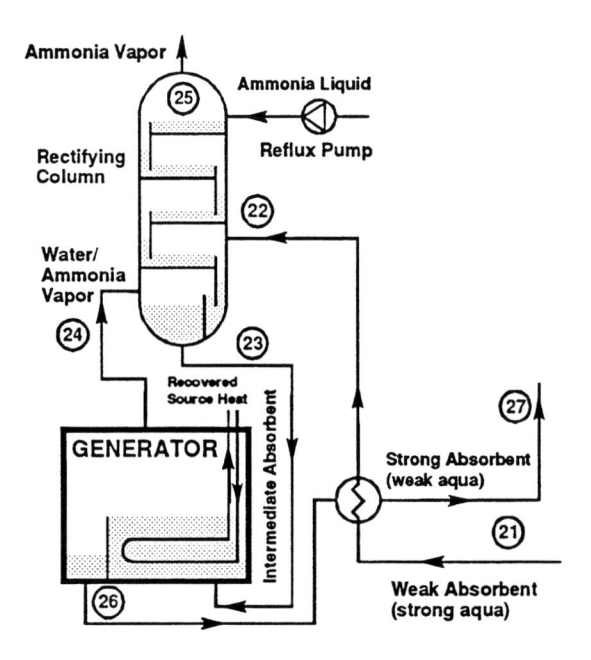

Figure 7-5 Rectification Column Detail

The strong absorbent (weak aqua) exits the generator at the bottom of the generator (point 26) and exchanges heat with the weak absorbent (strong aqua) through the solution heat exchanger.

7.4.6 Ammonia Condenser

The ammonia condenser can be one of several basic designs. The more popular systems are shell-and-tube heat exchangers and evaporative condensers. Regardless of the condenser used, there will be an ammonia receiver after the condenser to collect the liquid ammonia and serve as a surge drum.

7.4.7 Expansion Device

The expansion device in an absorption system provides a means to reduce the pressure of the liquid refrigerant and is the metering device that controls the flow of refrigerant to the evaporator(s). The expansion device for an AAR system is typically a hand expansion valve placed between the condenser and the evaporator.

The valve is adjusted to allow a specific flow of liquid refrigerant to leave the condenser/receiver at a specific depth of refrigerant in the condenser/receiver. As the level of refrigerant rises, the flow increases through the expansion valve. Conversely, as the flow decreases, the level falls. Figure 7-6 shows this process.

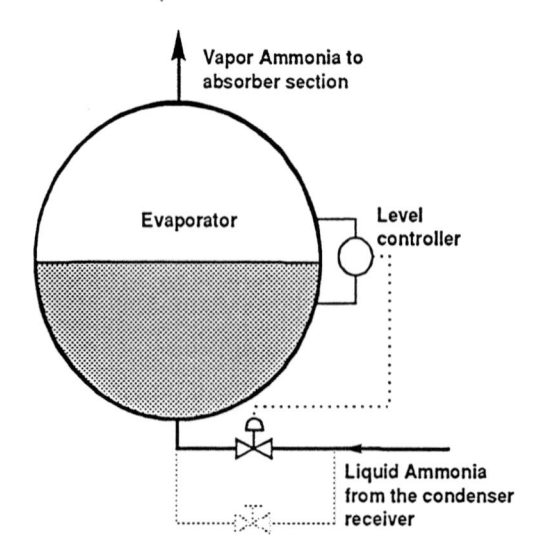

APPLICATION GUIDE FOR ABSORPTION COOLING/REFRIGERATION USING RECOVERED HEAT

Figure 7-6 Expansion Device Detail

7.4.8 Cooling Water System

The cooling water system is critical to the proper operation of an absorption machine. The heat generated in the absorber section and the heat of condensation in the condenser section must be removed and rejected to another medium, typically through an evaporative condenser or a cooling tower to the ambient outdoor air.

7.4.9 Controls

The primary purpose of the control system for an AAR machine is to control the flow of ammonia and absorbent through the machine in order to meet the cooling demand on the system. Meeting the demand is accomplished through a series of temperature and flow sensors that maintain pressures and flows in the absorber, condenser, generator, and evaporator. Essentially, as the demand on the machine decreases, the heat input to the generator also decreases.

7.5 CHEMICAL ADDITIVES

There are many different additives used in AAR systems. Similar to LiBr systems, the additives are used to protect the equipment from degradation and to improve the performance of the system. Currently, the AAR manufacturers are developing new additives that are easier to handle and are environmentally friendly. Since a transition is taking place on the use of additives, the individual manufacturers should be consulted to determine the proper additives to use and the benefits and drawbacks of the various additives available. The new additives should be considered for existing AAR machines.

REFERENCES

Shepherd, J. J. 1994. Personal communications (letter, fax, and phone).

BIBLIOGRAPHY

ASHRAE. 1988. *1988 ASHRAE handbook—Equipment*, p. 13.10. Atlanta: American Society of Heating, Refrigerating and Air-Conditioning Engineers, Inc.

ASHRAE. 1994. *1994 ASHRAE handbook—Refrigeration*, chapter 40. Atlanta: American Society of Heating, Refrigerating and Air-Conditioning Engineers, Inc.

Bogart, M. 1981. *Ammonia absorption refrigeration in industrial processes*. Houston: Gulf Publishing Co.

CHAPTER 8

Matching Heat Sources, Cooling Load, and Heat Rejection Capability with an Absorption Machine

8.1 INTRODUCTION

This chapter presents a detailed procedure for matching a heat source and a cooling load to an absorption machine. This procedure is described in generic terms. The application procedure includes the following general steps:

- determine cooling requirements and heat source,
- select type of absorption machine to use,
- calculate recoverable heat,
- determine the cooling output attainable,
- size the absorption machine,
- calculate heat rejection requirements,
- calculate electric power requirements,
- resize machine with other options, and
- determine part-load and nonconcurrent load operation.

The sections that follow are presented in an ideal sequence. In actual practice, the procedure may need to be approached differently, depending on the information available.

8.2 DETERMINE COOLING REQUIREMENTS AND HEAT SOURCE

The cooling or refrigeration load required and the design conditions must be determined first. The design conditions include

- leaving chilled-fluid temperature (°F [°C]);
- chilled-fluid temperature rise, difference at process (°F [°C]; this determines the chilled-water temperature entering the absorption machine);
- entering cooling water temperature (°F [°C]);
- cooling water temperature rise in the absorption machine (°F [°C]); and
- ambient wet-bulb temperature (°F [°C]).

In addition to the cooling requirements, the composition of the recoverable heat source must be determined to establish the quantity and quality of recoverable heat. Composition characteristics include temperature, flow rate, specific heat, and state (liquid or gas). The recovered heat can be from either a gas stream, as in the exhaust from a gas turbine, or from a liquid stream, as in a wastewater treatment facility. Recovered heat can also be used directly, such as cooling water from an engine coolant system or exhaust steam from a backpressure steam turbine.

8.3 SELECT TYPE OF ABSORPTION MACHINE TO USE

The type of absorption machine is typically chosen based on the cooling requirements and the available heat input temperature. Table 8-1 can be used as a guide for selecting the type of absorption machine based on these parameters.

TABLE 8-1
Selecting Absorption Machine Type

	Single-Effect LiBr Machine	Double-Effect LiBr Machine	Low-Temperature Ammonia Machine
Steam input, psig (kPa)	5–15 (135–205)	65–145 (110–120)	0–200 (100–1480)
Steam input, °F (°C)	225–250 (110–120)	305–365 (175–185)	210–385 (100–195)
Hot water input, °F (°C)	240–270 (115–130)	310–400 (155–205)	210–385 (100–195)
Cooling output, °F (°C)	40–60 (4–16)	40–60 (4–16)	–60 to 40 (–51 to 4)
COP	0.6–0.7	0.9–1.14	0.1–0.8

Note: This table is based on manufacturers' data available in 1994.

8.4 CALCULATE RECOVERABLE HEAT

The amount of recoverable heat can be calculated using the formulas presented in this section. For initial estimates, the calculations for determining the amount of recoverable heat are split into two primary groups based on the state of the source heat stream. The state is either gas or liquid. The state of the recovered heat stream must also be known. This state will usually be either gas (steam) or liquid (water).

If the source heat stream is used to generate steam, then two things must be considered. The steam should be superheated to overcome temperature losses, and the condensate should be subcooled to avoid flashing in the condensate lines.

8.4.1 Gas Streams

Gas streams have been applied directly to absorption machines as the heat source. The chemical composition of the gas stream must be analyzed and the individual equipment manufacturer contacted for any special requirements. The following sections assume that a heat exchanger will be used in order to isolate the systems. If no heat exchanger is used, then η_{eff} in Equation 8-1 would be 1.0 and $T_{mean,water}$ would be the leaving gas temperature. If steam is supplied directly to the absorption machine, as in an ebulliently cooled engine, then Equation 8-6 is solved for $q_{recovered}$.

Equation 8-1 can be used to estimate the amount of heat recoverable from exhaust gases. Equation 8-1 is applicable for both exhaust-gas-to-steam and exhaust-gas-to-water heat exchangers. This equation is based on available manufacturers' data.

$$q = \eta_{eff}(c_{p,gas})(m_{gas})\left(T_{in,gas} - T_{mean,water}\right) \quad (8\text{-}1)$$

where

q	=	heat recovered, Btu/h (kW$_T$);
η_{eff}	=	correction factor for water temperature, heat exchanger effectiveness, and leaving gas temperature (use values in Table 8-2);
$c_{p,gas}$	=	specific heat of source heat gas, ranges from 0.242 to 0.254 Btu/lb•°F (1.01 to 1.06 kJ$_T$/kg•°C);
m_{gas}	=	mass flow rate of source heat gas, lb/h (kg/s);
$T_{in,gas}$	=	temperature of heat source gas entering heat exchanger, °F (°C); and
$T_{mean,water}$	=	average temperature of recovered heat (water) in heat exchanger (or temperature of saturated steam), °F (°C).

The mean water temperature is based on the temperatures across the generator. Refer to chapter 9, Table 9-1, for guidelines on how to determine this value.

TABLE 8-2
Correction Factor for Exhaust Gas Heat Exchanger

Minimum Gas Flow lb/h (kg/s)	Maximum Gas Flow lb/h (kg/s)	Average η_{eff}
5000 (.063)	20,000 (2.52)	0.84
15,000 (1.89)	90,000 (11.34)	0.93

Note: The information in this table is based on manufacturers' published ratings. The correction factor, η_{eff} is a combination of factors for water temperature, heat exchanger effectiveness, and leaving gas temperature. The correction factor is relatively accurate with water temperatures above 200°F (90°C) and a leaving gas temperature above 300°F (150°C).

If specific information is not available for the heat source plant system, estimate heat recovery from gas turbines at 11 to 13 lb/h of steam (5 to 5.9 kg/hr) per kW of electricity generated. For reciprocating engines, exhaust gas heat recovery is approximately 1.3 to 1.4 lb/h of steam (0.59 to 0.64 kg/hr) recovered per kW of electricity generated. These estimates are reasonably accurate between 15 and 150 psig (200 and 1,100 kPa) (ASHRAE 1992).

The exhaust gas leaving temperature must be kept above 300°F (175°C) to avoid condensation of water vapor (ASHRAE 1992). Use Equation 8-2 to verify the exhaust gas leaving temperature when the heat recovered was estimated with Equation 8-1.

$$T_{gas,leaving} = T_{gas,entering} - \frac{q_{recovered}}{c_{p,gas} m_{gas}} \qquad (8-2)$$

where

$T_{gas,leaving}$ = temperature of heat source exhaust gas leaving heat exchanger (°F [°C]);
$T_{gas,entering}$ = temperature of heat source exhaust gas entering heat exchanger (°F [°C]);
$q_{recovered}$ = heat recovered (Btu/h [kW_T]);
$c_{p,gas}$ = specific heat of heat source gas, ranges from 0.242 to 0.254 Btu/lb·°F (1.01 to 1.06 kW·s/kg·°C); and
m_{gas} = mass flow rate of heat source gas (lb/h [kg/s]).

If the exhaust gas leaving temperature is determined to be below 300°F (149°C), recalculate the amount of recoverable heat assuming a leaving exhaust gas temperature of 300°F (149°C). Equation 8-2 can be rearranged for this purpose (see Equation 8-3).

$$q_{recovered} = \left(T_{gas,entering} - T_{gas,leaving}\right)\left(c_{p,gas}\right)\left(m_{gas}\right) \quad (8\text{-}3)$$

8.4.2 Liquid Streams

High-efficiency heat exchangers are available to transfer heat between two liquid streams. Equation 8-4 can be used to estimate the recoverable heat from a liquid stream.

$$q_{recovered} = c_{p,fl}\left(m_{fl}\right)\left(T_1 - T_2\right) \quad (8\text{-}4)$$

where

$q_{recovered}$ = heat recovered, Btu/h (kW$_T$);
$c_{p,fl}$ = specific heat of heat source fluid, 1.00 Btu/lb•°F (4.184 kW•s/kg•°C) for water;
m_{fl} = mass flow of heat source fluid, lb/h (kg/s);
T_1 = temperature of source (recovery) heat stream entering heat exchanger, °F (°C); and
T_2 = temperature of source (recovery) heat stream leaving heat exchanger, °F (°C).

When using Equation 8-4, the approach of the heat source stream temperature and the water temperature required in the absorption generator can be assumed to be 10°F (5.5°C) (Manufacturer 1994). Single-effect machines require an inlet temperature of between 170°F and 300°F (75°C and 150°C) water. Double-effect absorption units require an inlet temperature of between 270°F and 400°F (130°C and 200°C) water.

8.4.3 Engine Jackets

The amount of heat recoverable from the water jacket of a reciprocating engine for direct supply to the absorption machine can be estimated using Equation 8-5.

$$q_{jacket} = m_w c_{p,w}\left(T_1 - T_2\right) \quad (8\text{-}5)$$

where

q_{jacket} = heat recovered, Btu/h (kW$_T$);
m_w = mass flow of water from engine jacket in lb/h (kg/s);
$c_{p,w}$ = specific heat of water, 1.00 Btu/lb•°F (4.184 kW•s/kg•°C);

T_1 = temperature out of engine, °F (°C); and
T_2 = temperature into engine, °F (°C).

Based on manufacturers' published material, the water temperature rise across an engine jacket, $T_1 - T_2$, is approximately 50°F (28°C) (Manufacturer 1994).

8.4.4 Backpressure Steam Turbine

Backpressure steam turbines can be obtained to produce steam pressures applicable for absorption cooling. Table 8-3 gives typical steam rates for a multistage steam turbine (Smith et al. 1992).

TABLE 8-3
Multistage Steam Turbine Generator Steam Rates

Turbine Inlet Conditions		Turbine Outlet Steam Rates, lb/h/kW (kJ/s•kW)	
Steam Pressure psig (kPa)	Temperature °F (°C)	with 12 psig steam backpressure (185 kPa)	with 115 psig steam backpressure (894 kPa)
150 (1100)	366 (185)	257 ...	... ...
350 (2500)	500 (260)	18 (5)	55 (16)
400 (2900)	650 (340)	14 (4)	29 (8)
850 (6000)	825 (440)	10 (3)	17 (5)
1250 (8700)	900 (480)	9 (3)	14 (4)

8.4.5 Mass Flow of Recovered Heat

Once the amount of heat recovered from a source heat stream has been calculated, the mass flow of the recovered heat stream can be calculated. If steam is the final medium of the recovered heat, then Equation 8-6 should be used. In determining the enthalpy difference of the steam in Equation 8-6, the amount of superheat and subcool must be accounted for. The required superheating and subcooling for different systems vary and are therefore discussed in later chapters.

$$m_s = \frac{q_{recovered}}{\left(h_g - h_f\right) K_5} \tag{8-6}$$

where

m_s	=	mass flow rate of steam, lb/h (kg/s);
$q_{recovered}$	=	heat recovered from jacket/exhaust gas, Btu/h (kW$_T$);
h_g	=	enthalpy of saturated steam at desired pressure, Btu/lb (kW$_T$•s/kg);
h_f	=	enthalpy of saturated condensate at desired pressure, Btu/lb (kW$_T$•s/kg); and
K_5	=	correction factor for superheating and subcooling, %/100.

If water is the final medium of the recovered heat, then Equation 8-7 can be used. The temperature increase of the recovered heat stream is based on the requirements of the individual absorption machine. The values are discussed in later chapters.

$$m_w = \frac{q_{recovered}}{c_{p,w} \left(T_o - T_i\right)} \tag{8-7}$$

where

m_w	=	mass flow rate of water, lb/h (kg/s);
$q_{recovered}$	=	heat recovered from jacket/exhaust gas, Btu/h (kW$_T$);
$c_{p,w}$	=	specific heat of water, 1.00 Btu/lb•°F (4.184 kW$_T$•s/kg•°C);
T_o	=	temperature of water out, °F (°C); and
T_i	=	temperature of water in, °F (°C).

8.5 DETERMINE THE COOLING OUTPUT ATTAINABLE

Once the recovered heat properties are known, calculate the amount of cooling available from the absorption machine using Equation 2-1. Equation 2-1 is based on the coefficient of performance (COP) of the absorption machine and is repeated below as Equation 8-8.

$$q_{cooling} = \frac{COP_{ABS}\left(q_{recovered}\right)}{C_1} \qquad (8\text{-}8)$$

where

$q_{cooling}$	=	cooling output, tons (kW_T);
COP_{ABS}	=	coefficient of performance of absorption machine;
$q_{recovered}$	=	heat recovered from jacket/exhaust gas, Btu/h (kW_T); and
C_1	=	conversion factor, 12,000 Btuh/ton (kW_T/kW_T).

The cooling output is the amount of cooling provided under specific conditions of the system and not the nominal size of the machine. The COPs for various types of absorption machines are included in other chapters, including chapters 9 and 10.

8.6   SIZE THE ABSORPTION MACHINE

The nominal size of an absorption machine is typically based on ARI conditions of 85°F (29°C) entering cooling water temperature and a 44°F (7°C) leaving chilled-water temperature (ARI 1992). Also, an inlet steam pressure is used to define the nominal capacity of the absorption machine.

If site-specific operating conditions are different than nominal conditions, then the cooling output must be corrected to determine nominal conditions. The nominal size of the machine is determined in order to easily compare different systems, manufacturers, and conditions. Correction factors for a different steam pressure and a different leaving chilled-water temperature can be determined given the input/output temperatures and pressures.

The nominal size of an absorption machine does not vary significantly with changes in the chilled-water temperature difference. The nominal machine size is based on the leaving chilled-water temperature. As the design entering chilled-water temperature changes, the flow rate changes proportionally.

8.7   CALCULATE HEAT REJECTION REQUIREMENTS

All absorption machines must reject the heat produced in the absorber section and the heat of condensation from the condenser section. Typically, heat rejection is accomplished with a cooling tower or evaporative condenser; however, other methods, such as boiler feed water preheating, can also be considered. Tables or equations can be generated or manufacturers' data can be obtained to determine the heat rejection requirements for an absorption machine.

8.8 CALCULATE ELECTRIC POWER REQUIREMENTS

The primary direct electrical use of an absorption machine is that of the solution/refrigerant pump(s). The cooling water pump and fan energy use will be significant and must also be determined.

8.9 RESIZE MACHINE WITH OTHER OPTIONS

If alternative heat sources are available, cooling output temperatures can be varied to meet alternative loads, or changes are desired due to maintenance considerations or owner preference alternatives, then the procedure above should be repeated for each combination to determine the best option available. One instance where resizing is advisable is when a heat source could produce either steam or high-temperature hot water. This difference in the state of the heat source could affect the sizing and type of absorption machine selected. Resizing should also be accomplished to determine the effect of varying cooling water temperatures on the absorption machine size and heat rejection requirements.

8.10 DETERMINE PART-LOAD AND NONCONCURRENT LOAD OPERATION

With an absorption system, contingencies must be available when the amount of recovered heat and the cooling loads are not matched with the absorption machine. The following scenarios are examples: excess recovered heat available in relation to cooling load and insufficient recovered heat available in relation to cooling load.

8.10.1 Excess Recovered Heat

If there is excess recovered heat available, four options are available:

- transfer heat to storage for later use in absorption cooling,
- transfer heat to another heating load if available,
- reject the heat through a cooling tower, and
- operate absorption machine at full cooling capacity and store the excess cooling.

Rejection of the excess recovered heat to a cooling tower achieves the desired result, but the recovered heat is not fully utilized. Another option, if possible, is to store the heat in a thermal storage system for later use by the absorption machine or for another process. Heat storage is covered in detail in other publications (Lenard 1994; Tamblyn 1985; ASHRAE 1991).

A similar option is to store the excess cooling provided by a fully loaded absorption machine using a cool thermal storage system. The best cool thermal storage option for the majority of absorption systems is a stratified chilled-water storage system at 40°F (4°C) or eutectic salt systems at 48°F (9°C). Chilled-water storage may also be used as a fire protection tank, which reduces the project's first cost. With ammonia absorption systems, any number of the available ice storage systems are also good options. A complete discussion of cool thermal storage applications is contained in other publications (Dorgan and Elleson 1993).

8.10.2 Insufficient Recovered Heat

If there is insufficient recovered heat available to produce the required cooling, several options are available:

- install and use cooling from thermal storage system,
- install a supplementary boiler to produce heat for the absorption system, and
- install supplementary cooling equipment.

If a cool thermal storage system is available, either heat or cool storage, then the energy stored during low cooling demand periods can be used when not enough recovered heat is available for the absorption system to meet the entire cooling load. The thermal storage option will increase the first cost of the system.

A second option is to install a supplementary boiler to produce additional heat required by the absorption machine. Although this solves the cooling needs, a supplementary boiler increases first costs, fuel use, and maintenance costs. However, for short periods of use, a supplementary boiler may be an economical first-cost option versus supplementary cooling machines.

A third option is to install a vapor-compression machine for supplementary cooling. A vapor-compression system increases first, operating, and fuel costs. However, if a vapor-compression system is already installed, this can be an economical alternative.

8.11 AUTOMATION OF APPLICATION PROCEDURE

Most of the manufacturers of absorption machines have computer software that aids in the selection and application of absorption machines for a specific site. For any detailed analysis, use these programs in combination with the information in this guide.

For initial investigations into general absorption applications, a computer program (ASHSORPT) has been included with this guide. The algorithms used in the computer program are contained in Appendix B.

The input variables to ASHSORPT is the heat stream to be recovered and cooling output conditions. The output of ASHSORPT includes the estimated nominal machine size and heat rejection requirement for the input provided. A detailed discussion of the computer program is contained in Appendix C.

REFERENCES

ARI. 1992. *Standard 560, Absorption water chilling and water heating packages.* Arlington, VA: Air-Conditioning and Refrigeration Institute.

ASHRAE. 1991. *1991 ASHRAE handbook—HVAC applications*, chapter 39. Atlanta: American Society of Heating, Refrigerating and Air-Conditioning Engineers, Inc.

ASHRAE. 1992. *1992 ASHRAE handbook—HVAC systems and equipment*, chapter 7. Atlanta: American Society of Heating, Refrigerating and Air-Conditioning Engineers, Inc.

Dorgan, C. E., and J. S. Elleson. 1993. *Design guide for cool thermal storage.* Atlanta: American Society of Heating, Refrigerating and Air-Conditioning Engineers, Inc.

Lenard, J. 1994. Thermal energy storage increases Cogen abilities. *ASME News*, February, p. 5.

Manufacturer. 1994. Personal communication (phone). Bell & Gossett.

Smith, W. P., Jr., R. C. Erickson, W. A. Liegois, and C. E. Dorgan. 1992. *Cogeneration technology.* Course text, College of Engineering, EPD Department. Madison: University of Wisconsin.

Tamblyn, R. T. 1985. College Park thermal storage experience. *ASHRAE Transactions* 91(1B): 947–955.

CHAPTER 9
LiBr Absorption Machine Applications

9.1 INTRODUCTION

This chapter applies the procedure from chapter 8 to LiBr absorption machines. The charts and associated algorithms developed in this chapter were obtained by modeling the performance of typically available absorption machines. With this approach, the information contained in this guide generates a reliable first-level estimate for the application of LiBr absorption machines. Since the charts and algorithms are not for a specific machine, a detailed analysis must be accomplished using manufacturers' data for the actual design of an absorption system. Appendix B contains a complete discussion of the development of the algorithms and lists all of the algorithms used for the charts in this chapter.

Examples have been included to show the wide range of applications covered by the procedure developed in this guide. The primary example used is similar in size and configuration to the sites listed in Appendix D. This example also is used to present the charts and algorithms developed for LiBr absorption systems.

As most operating and design conditions are not at ARI conditions, the design cooling requirements must be translated to ARI nominal conditions in order to easily compare different system options. The procedure from chapter 8 determines the nominal ARI capacity given non-ARI design conditions. The examples contained in this chapter have non-ARI conditions in order to demonstrate the procedure from chapter 8.

9.2 EXAMPLE DESCRIPTION

The example presented is for an industrial plant located in central Massachusetts. A gas turbine is used to generate electricity for the plant. Recovered heat is obtained from the exhaust stack of the gas turbine. Located near the industrial plant is an office complex with a year-round average cooling load of approximately 600 tons (2,110 kW$_T$) and a peak summer cooling load of approximately 1,500 tons (5,375 kW$_T$). There is also a year-round process cooling load of 400 tons (1,410 kW$_T$) required by the industrial plant. The heat rejection medium for the absorption machine is a cooling tower. The layout for the example absorption system is shown in Figure 9-1.

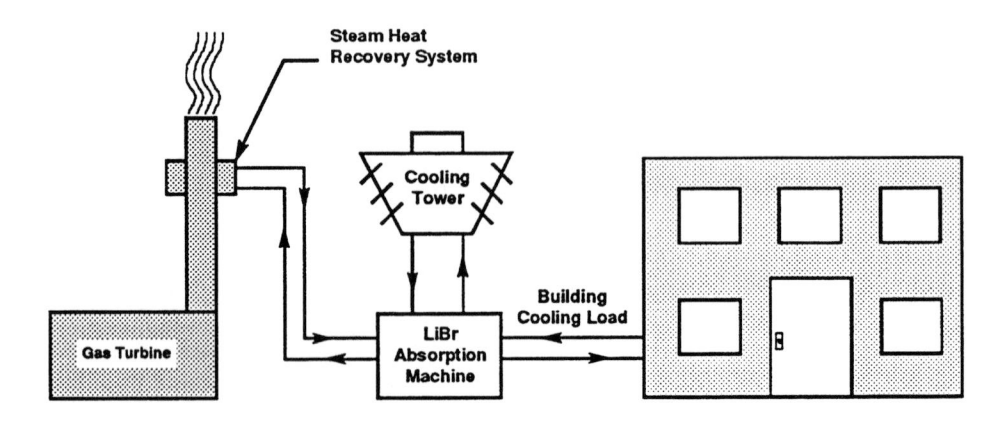

Figure 9-1 Layout for LiBr Example

For central Massachusetts, the outdoor design wet-bulb temperature is approximately 73°F (23°C). With a 7°F (4°C) approach on the cooling tower, this equates to an entering cooling water temperature to the absorption machine of 80°F (27°C). A temperature rise of 10°F (6°C) across the absorption machine equates to a 90°F (32°C) leaving cooling water temperature. The chilled-water entering temperature is 60°F (15°C) and the chilled-water leaving temperature is 42°F (6°C).

The gas turbine is a 1-MW unit. At ambient conditions of 80°F (27°C), the exhaust gas of the gas turbine is approximately 960°F (516°C) at 61,000 pounds per hour (7.9 kg/s) (Manufacturer 1994).

9.3 DETERMINING COOLING REQUIREMENTS AND HEAT SOURCE

The first step in applying an absorption system is to determine the cooling requirements and design conditions. The following values were obtained from the problems statement:

- leaving chilled-water temperature = 42°F (6°C),
- chilled-water temperature rise = 18°F (10°C),
- entering cooling water temperature = 80°F (27°C),
- cooling water temperature rise = 10°F (6°C), and
- ambient wet-bulb temperature = 73°F (23°C).

LiBr ABSORPTION MACHINE APPLICATIONS

These values are also from the problem statement:

- entering exhaust gas temperature = 960°F (516°C);
- flow rate of exhaust gas = 61,000 lb/h (7.7 kg/s); and
- state of heat source = gas.

Equation 8-1 from chapter 8 stated that the exhaust gas specific heat ranges from 0.242 to 0.254 Btu/lb•°F (1.01 to 1.06 kJ$_T$/kg•°C). A value of 0.25 Btu/lb•°F (1.05 kJ$_T$/kg•°C) is a good estimate to use if the exhaust gas specific heat is unknown.

9.4 SELECT TYPE OF ABSORPTION MACHINE TO USE

The second step is to determine what type of absorption machine to use. The information in Table 8-1 from chapter 8 shows that either a single-effect or a double-effect absorption machine is suitable for this application. Since the temperature of the exhaust gas is a higher-end medium-grade heat, a double-effect machine is likely to be the most efficient for this application.

Figure 9-2 contains information on steam pressure and temperature and is based on an absorption machine with constant output at an inlet steam pressure of 12 psig (184 kPa) for single-effect machines and 115 psig (894 kPa) steam pressure for double-effect machines. Chilled water at 44°F (7°C) is the output of the machine. If the exhaust stream is to be used directly, then use the temperature axis as a guide.

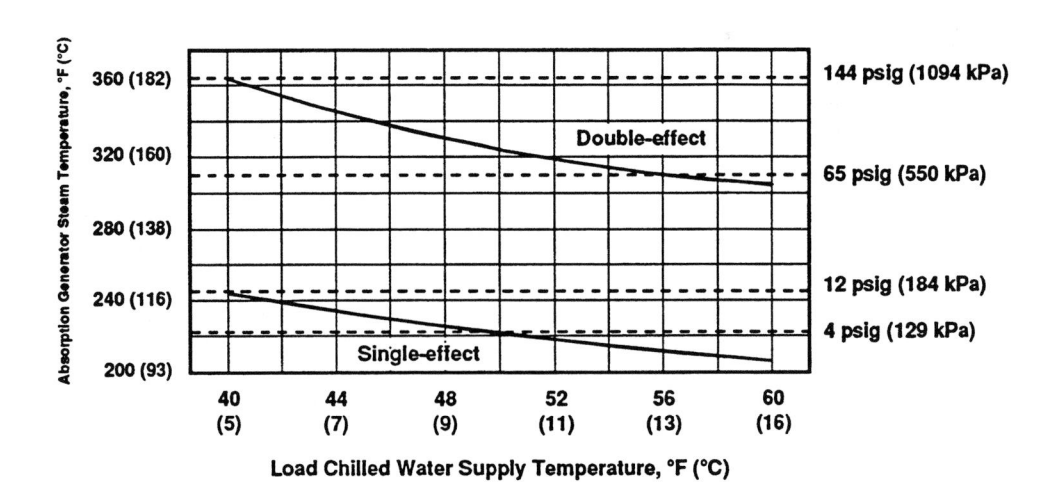

Figure 9-2 LiBr Absorption Machine Input Temperatures vs. Output Temperatures

109

Figure 9-2 shows that to obtain lower temperature chilled water, a higher quality (pressure and temperature) steam is required. The steam pressure for a specific application should fall between the thick dashed horizontal lines in Figure 9-2. For the initial application, choose the higher steam pressure. For applications where recovered hot water is available in lieu of steam, Table 9-1 contains information for determining the equivalent steam pressure given the hot water entering temperature to the absorption machine and the temperature decrease through the absorption machine. The equivalent steam pressure obtained from Table 9-1 is used in subsequent calculations whenever steam pressure is required for a specific formula or selection capacity. The information in Table 9-1 is based on manufacturers' data (Manufacturers 1994). Table 9-1 provides an example application.

Since the exhaust gas temperature from the gas turbine in this example is 960°F (516°C), steam at any of the temperatures shown in Figure 9-2 can be generated to power this absorption system. A pressure of 115 psig (894 kPa) is a midpoint of machine operation and is a good starting point. From a reference steam table (such as Table A-2, Appendix A), the saturated temperature of steam at 115 psig (894 kPa) is 347.1°F (175.1°C).

9.5 CALCULATE RECOVERABLE HEAT

The third step is to calculate the amount of recoverable heat attainable from the heat source. As this example is a gas-to-gas (steam) heat recovery system, Equation 8-1 (gas stream heat recovery) from chapter 8 can be used to determine the quantity of heat recovered. Equation 8-1 becomes

$$\begin{aligned} q &= \eta_{eff}(c_{p,gas})(m_{gas})\left(T_{in,gas} - T_{mean,water}\right) \\ &= 0.93\,(0.25)\,(61,000)\,(960 - 347.1) = 8,692,454\,Btuh \\ &= 0.93\,(1.05)\,(7.7)\,(516 - 175.1) = 2563\,kW_T \end{aligned} \quad (9\text{-}1)$$

The mean temperature of the water in this example is the saturated steam temperature at the chosen pressure. Saturated steam temperatures, pressures, and enthalpies are contained in Appendix A or a standard steam properties table.

Before converting the result from Equation 9-1 to a steam flow rate using Equation 8-6, the effect of superheating and subcooling of the steam must be accounted for. Superheating of the steam ensures that it does not condense as its temperature is lowered as it passes through the piping system. Subcooling of the steam is desirable so the condensate does not flash to steam in the return lines. Subcooling will minimize line sizes for nonpumped systems. Heat recovery from the steam condensate return is desirable whenever there is a use for the heat.

The values for K_5 in Equation 8-6 (mass flow rate of recovered heat) are based on recommendations from heat recovery and absorption equipment manufacturers. These values are

TABLE 9-1
Equivalent Steam Pressure for Hot Water, psig (kPa)

Inlet Temp., °F (°C)	210 (99)	220 (104)	230 (110)	240 (116)	290 (143)	300 (149)	310 (154)	320 (160)	330 (166)	340 (171)
Single-Effect Machine										
240 (116)	4.1 (130)	6.4 (146)	9.0 (163)	...	...	...	...	...	...	...
250 (121)	5.4 (139)	7.9 (156)	10.2 (172)	13.0 (194)	...	...	...	...	...	...
260 (127)	5.9 (142)	8.3 (159)	10.5 (174)	14.0 (200)	...	...	...	...	...	...
270 (132)	6.4 (146)	8.5 (160)	10.7 (175)	14.0 (200)	...	...	...	...	...	...
Double-Effect Machine										
310 (154)	...	...	...	...	43 (398)	...	...	...	...	...
320 (160)	...	...	...	...	46 (419)	55 (481)	...	...	...	...
330 (166)	...	...	...	...	50 (446)	57 (494)	65 (550)	...	...	...
340 (171)	...	...	...	...	52 (460)	59 (508)	68 (570)	75 (618)	...	...
350 (177)	...	...	...	...	54 (474)	60 (515)	70 (584)	78 (639)	88 (708)	...
360 (182)	...	...	...	...	56 (487)	62 (529)	72 (598)	80 (653)	93 (743)	...
370 (188)	...	...	...	...	58 (501)	64 (543)	73 (605)	83 (674)	97 (770)	115 (894)
380 (193)	...	...	...	...	59 (508)	65 (550)	75 (618)	86 (694)	100 (791)	120 (929)
390 (199)	...	...	...	...	60 (515)	67 (563)	76 (625)	88 (708)	103 (812)	...
400 (204)	...	...	...	...	62 (529)	68 (570)	78 (639)	90 (722)	106 (832)	...

Outlet Temperature, °F (°C)

Example:

Inlet: 360°F (182°C)
Outlet: 320°F (160°C)
Equivalent steam: 80 psig (653 kPa)

where psig (approximate) = psia (values in Table A-1, Appendix A) − 15 (the kPa values listed in Table 9-1 are the same as those in Table A-1).

- single-effect machine: $K_5 = 1.03$,
- double-effect machine: $K_5 = 1.23$, and
- direct steam use: $K_5 = 1.00$ (solve Equation 8-6 for $q_{recovered}$).

Therefore, Equation 8-6 becomes

$$m_s = \frac{q_{recovered}}{(h_g - h_j)(K_5)}$$

$$= \frac{8,692,454}{(1191.7 - 318.6)(1.23)} = 8,094\ lb/hr\ of\ 115\ psig\ steam \quad (9\text{-}2)$$

$$= \frac{2563}{(2771.9 - 741.1)(1.23)} = 1.03\ kg/s\ of\ 894\ kPa\ steam$$

The enthalpies in Equation 9-2 are the saturated enthalpies of the input steam and leaving condensate at the chosen pressure of the absorption system. Appendix A contains tables of these values.

To ensure that condensation does not form in the exhaust gas stack of the gas turbine, the exhaust gas temperature must be kept above 300°F (150°C). To calculate the leaving exhaust gas temperature from the heat recovery system, use Equation 8-2 (exhaust gas leaving temperture). For the example, Equation 8-2 becomes

$$T_{gas,leaving} = T_{gas,entering} - \frac{q_{recovered}}{c_{p,gas} m_{gas}}$$

$$= 960 - \frac{8,692,454}{0.25\,(61,000)} = 390°F \quad (9\text{-}3)$$

$$= 516 - \frac{2,563}{1.05\,(7.7)} = 199°C$$

Since the leaving temperature is above 300°F (150°C), condensation will not occur in the stack. Approximately 8,000 pounds per hour (1.03 kg/s) of steam can be recovered from the exhaust stack of the gas turbine and provided to the generator of the double-effect absorption machine.

9.6 DETERMINE AMOUNT OF COOLING OUTPUT

The fourth step in the procedure is to determine the amount of cooling output attainable from a LiBr absorption machine with the given heat input. To simplify this process,

Figure 9-3 has been developed. Figure 9-3 is a graphical representation of the characteristics of a number of absorption machines currently available. The upper and lower boundaries in Figure 9-3 are the high and low steam pressure limits for the economic operation of a typical absorption system. These limits are

- single-effect machine:
  a = 12 psig (184 kPa)
  b = 4 psig (129 kPa)

- double-effect machine:
  c = 144 psig (1,099 kPa)
  d = 65 psig (550 kPa).

If the heat input is greater than the maximum given value in Figure 9-3, then multiple absorption machines are required.

For the example, Figure 9-3 indicates that for a heat input of 8.7×10^6 Btuh (2.54 MW_T), a double-effect absorption machine can economically provide between 740 and 895 tons (2,600 and 3,150 kW_T). Since this example uses 115 psig (894 kPa) of steam, which is somewhere between the two lines, the resultant estimated value for cooling with a recovered heat stream is approximately 800 tons (2,810 kW_T).

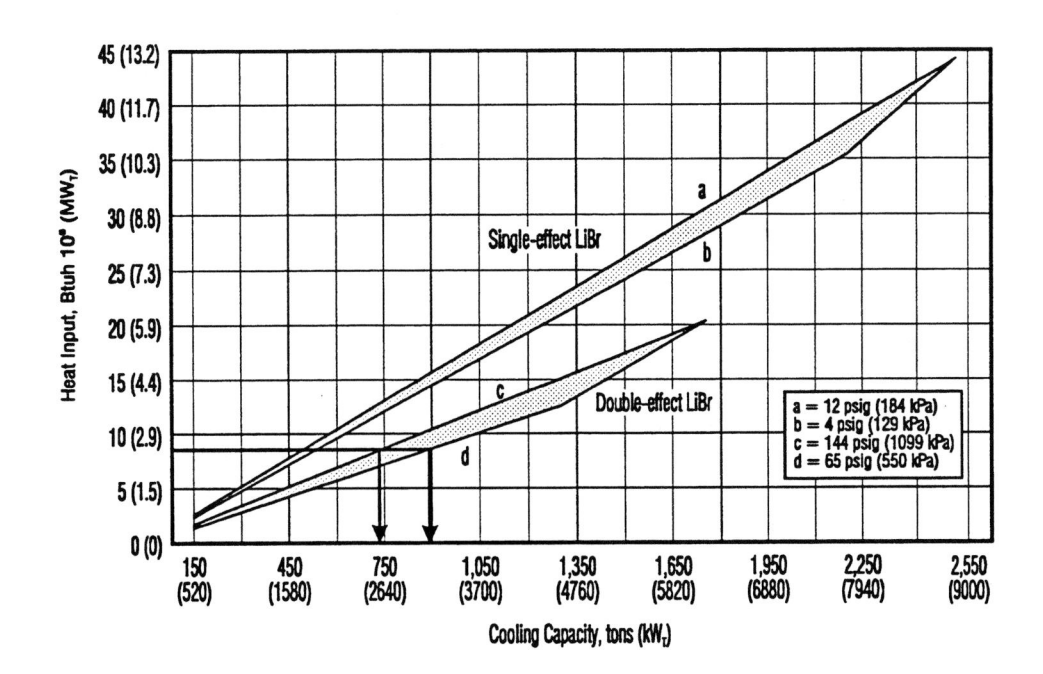

Figure 9-3 LiBr Input vs. Output Capacities

9.7 SIZE THE ABSORPTION MACHINE

Since machine capacities are published at ARI conditions, the required capacity determined in section 9.6 must be converted from actual design conditions to ARI conditions. The result is the nominal (ARI) capacity of the machine.

Therefore, the fifth step in the procedure is to determine the nominal size of the absorption machine. The nominal size is required to easily compare the various input and output options for a specific application. To determine the nominal size of the absorption machine, the estimated cooling output must be corrected for nonstandard steam pressures, cooling water temperatures, and chilled-water temperatures.

9.7.1 Steam Pressure Correction

To simplify correcting for steam pressures at other than the rated ARI conditions, Figures 9-4 and 9-5 have been developed.

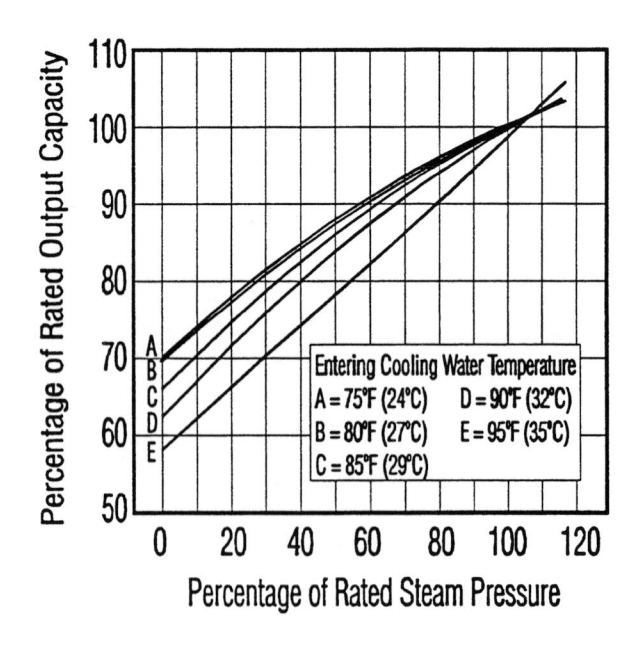

Note: 100% Rated Steam Pressure = 12 psig (184 kPa) Steam

Figure 9-4 Steam Pressure Derating Chart, Single-Effect

LiBr Absorption Machine Applications

Figure 9-4 is for a single-effect LiBr absorption machine with a rated steam pressure of 12 psig (184 kPa). Figure 9-5 is for a double-effect LiBr machine with a rated steam pressure of 115 psig (894 kPa). For the example, Figure 9-5 indicates that, for a steam pressure of 115 psig (894 kPa)—which is 100%—a double-effect absorption machine with an 80°F (27°C) entering cooling water temperature, the percent of rated output capacity is 100%. With this information, Equation 9-4 is used to determine the corrected capacity for steam pressure:

$$q_{cap-cor-s} = \frac{q_c}{K_6} \qquad (9\text{-}4)$$

where

$q_{cap-cor-s}$ = cooling capacity at rated steam pressure (ARI conditions), tons (kW$_T$);
q_c = cooling capacity value from Figure 9-3, tons (kW$_T$); and
K_6 = correction factor from Figure 9-4 or 9-5 divided by 100.

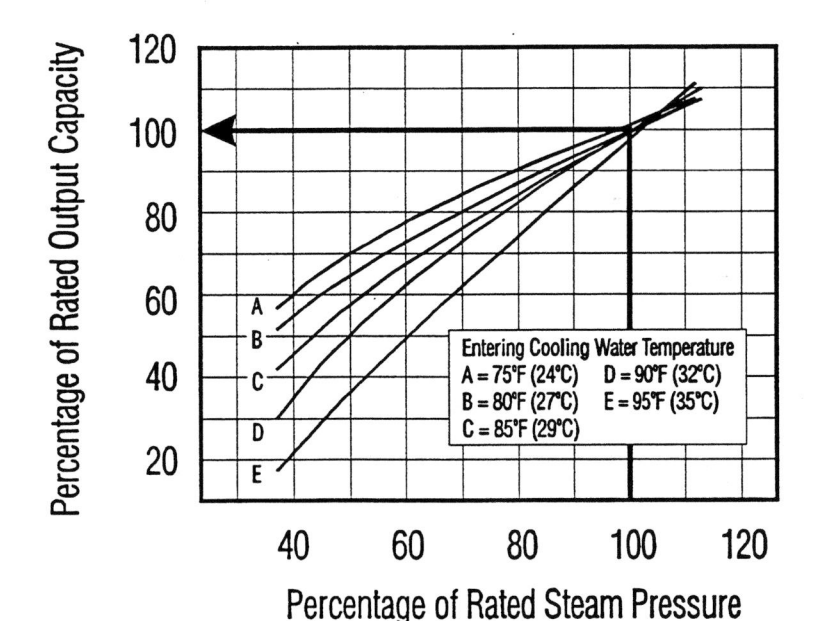

Figure 9-5 Steam Pressure Derating Chart, Double-Effect

Note: 100% Rated Steam Pressure = 115 psig (894 kPa)

Entering Cooling Water Temperature
A = 75°F (24°C)   D = 90°F (32°C)
B = 80°F (27°C)   E = 95°F (35°C)
C = 85°F (29°C)

APPLICATION GUIDE FOR ABSORPTION COOLING/REFRIGERATION USING RECOVERED HEAT

For the example,

$$q_{cap-cor-s} = \frac{800}{1.0} = 800\, tons \quad (manufacturer\ listed\ capacity)$$
$$= \frac{2810}{1.0} = 2810\, kW_T \quad (9\text{-}5)$$

9.7.2 Cooling Water and Chilled-Water Temperature Correction

The second correction to the cooling output accounts for variations in the entering cooling water temperature and the leaving chilled-water temperature. Figure 9-6 has been developed for single-effect machines and Figure 9-7 has been developed for double-effect machines.

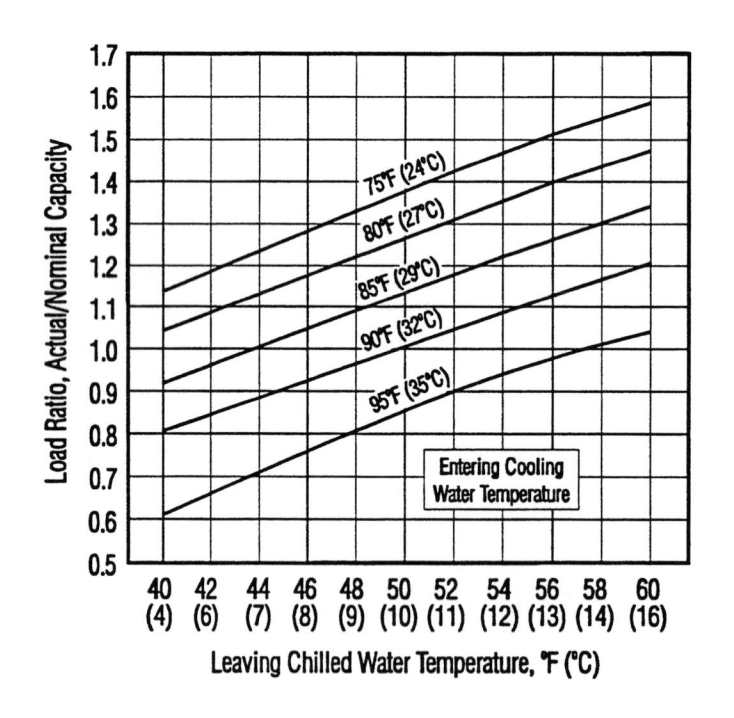

Figure 9-6 CHW and CW Correction Factors, Single-Effect

Figures 9-6 and 9-7 are based on a cooling water temperature rise of 10°F (6°C) through the absorption machine and a nominal steam pressure rating of 12 psig (184 kPa) for single-effect LiBr machines and 115 psig (894 kPa) for double-effect LiBr machines. At these conditions, the figures can be used directly.

LiBr ABSORPTION MACHINE APPLICATIONS

If a different cooling water temperature rise is required, the figures can still be used. For example, if the cooling water temperature rise is 15°F (8°C) and the entering cooling water temperature is 80°F (27°C), then in Figure 9-7, the 85°F (29°C) line would be used instead of the 80°F (27°C) line. The 85°F (29°C) line is used since the cooling water temperature leaving the absorption machine is 95°F (35°C), [80°F + (15°F − 10°F) = 95°F (27°C + (8°C − 6°C) = 29°C)]. This process results in a close approximation of the actual system performance (Plzak 1994).

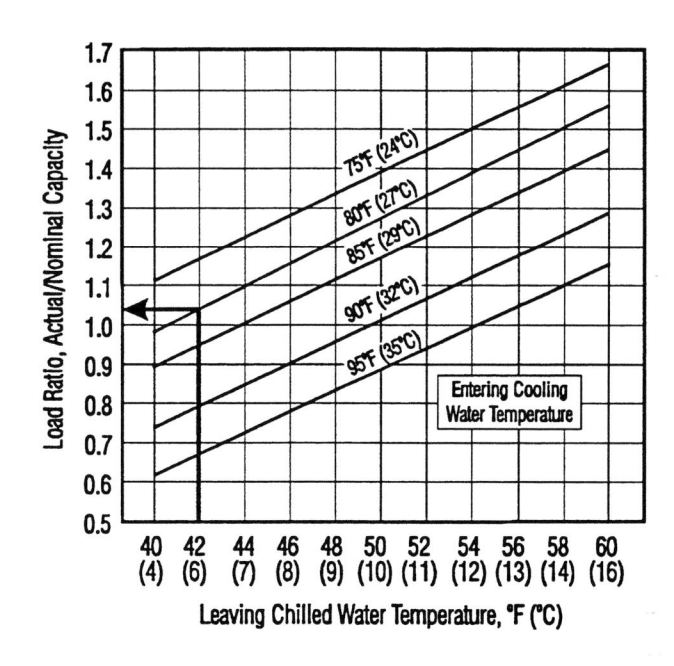

Figure 9-7 CHW and CW Correction Factors, Double-Effect

For the example, Figure 9-7 can be used directly and indicates that for a leaving chilled-water temperature of 42°F (6°C) and an entering cooling water temperature of 80°F (27°C), the load ratio is 1.05.

With the load ratio determined, Equation 9-6 can be used to calculate the corrected nominal rated value based on cooling output due to variations in the chilled-water and cooling water temperatures. This results in the nominal cooling capacity at ARI conditions of the absorption machine:

$$q_{cap-cor-T} = q_{cap-nom} = \frac{q_{cap-cor-s}}{K_7} \tag{9-6}$$

where

$q_{cap-cor-T}$ = cooling capacity at rated steam pressure and cooling temperature, tons (kW$_T$);
$q_{cap-nom}$ = nominal cooling capacity, tons (kW$_T$);
$q_{cap-cor-s}$ = from Equation 9-4, tons (kW$_T$); and
K_7 = correction factor from Figure 9-6 or 9-7, load ratio.

For the example,

$$q_{cap-nom} = \frac{800}{1.05} = 762 \, tons \quad (nominal\,ARI\,rated\,capacity)$$
$$= \frac{2810}{1.05} = 2670 \, kW_T \tag{9-7}$$

9.8　HEAT REJECTION REQUIREMENTS

The sixth step in the procedure is to determine the heat rejection requirements of the absorption machine. The variables in determining the heat rejection of an absorption machine are cooling water temperature rise through the absorption machine and type of absorption machine (single-effect versus double-effect). For an initial approximation, use Equation 9-8:

$$q_{heat-rej} = C_3 \left[q_c \bullet K_8 \right] \Delta T_{CT} \tag{9-8}$$

where

$q_{heat-rej}$ = heat rejection requirement, Btuh (kW$_T$);
C_3 = constant, 500 Btu•min/h•°F•gal (4.17 kJ$_T$•s/s•°C•L);
q_c = cooling output at design conditions from Figure 9-3, tons (kW$_T$);
K_8 = gpm/ton (L/s•kW$_T$); and
ΔT_{CT} = temperature difference across cooling tower, °F (°C).

The term $C_3 \bullet K_8 \bullet \Delta T_{CT}$ can be considered a constant for initial estimations. Therefore, Equation 9-8 becomes

$$q_{heat-rej} = C_4 \cdot q_c \qquad (9\text{-}9)$$

where

$q_{heat-rej}$ = heat rejection requirement, Btuh (kW$_T$);
C_4 = constant, single-effect = 30,600 Btuh/ton (2.54 kW$_T$/kW$_T$)
double-effect = 22,500 Btuh/ton (1.87 kW$_T$/kW$_T$); and
q_c = cooling output at design conditions from Figure 9-3, tons (kW$_T$).

For the example,

$$\begin{aligned} q_{heat-rej} &= 22,500 \cdot 800 = 18,000,000\, Btuh \\ &= 1.87 \cdot 2810 = 5236\, kW_T \end{aligned} \qquad (9\text{-}10)$$

9.9 ELECTRICAL POWER REQUIREMENTS

The seventh step is to estimate the electrical power requirements for the absorption system. This includes determining the electrical demand and use of the absorption machine pumps, cooling water pumps, and cooling tower fans. The information, originally presented in chapter 3, is summarized below.

9.9.1 Absorption Machine Electrical Energy Use

The primary electrical energy consumed by the absorption machine is used to power the refrigerant and solution pumps. A listing of the electrical use for LiBr absorption machines is contained in Table 9-2.

TABLE 9-2
Absorption Machine Electrical Energy Use

Machine Type	Machine Size, tons (kW$_T$)	Electrical Use, kW/ton (kW/kW$_T$)	Recommended Value kW/ton (kW/kW$_T$)
Single-Effect	100–200 (350–700)	0.014–0.055 (0.004–0.016)	0.030 (0.009)
	200–400 (700–1400)	0.011–0.028 (0.003–0.008)	0.022 (0.006)
	400–600 (1400–2100)	0.010–0.019 (0.003–0.005)	0.015 (0.004)
	600–1,000 (2100–3500)	0.008–0.015 (0.002–0.004)	0.012 (0.003)
	1,000–1,600 (3500–5600)	0.006–0.017 (0.002–0.005)	0.012 (0.003)
Double-Effect	100–250 (350–900)	0.019–0.023 (0.005–0.007)	0.021 (0.006)
	250–450 (900–1600)	0.015–0.028 (0.004–0.008)	0.020 (0.006)
	450–800 (1600–2800)	0.011–0.027 (0.003–0.008)	0.018 (0.005)
	800–1,500 (2800–5300)	0.009–0.026 (0.003–0.007)	0.015 (0.004)

Notes: The tons (kW$_T$) listed are for a nominal-sized machine at ARI conditions (44°F [7°C] chilled water, 85°F [29°C] cooling water, 12 psig [184 kPa] steam single-effect, and 115 psig [894 kPa] steam double-effect).

The kW/ton (kW/kW$_T$) ranges are a compilation of those of all major manufacturers. The recommended kW/ton (kW/kW$_T$) is a weighted average and is a good approximation for average use.

With the information from Table 9-2, Equation 3-1 from chapter 3 can be used to determine the absorption machine energy use. For the example,

$$AM_{kWh} = K_1 \cdot q_{cap-nom} \cdot H_1$$
$$= 0.018 \cdot 762 \cdot 8,736 = 120000 \, kWh$$
$$= 0.005 \cdot 2670 \cdot 8736 = 117000 \, kWh \quad (9\text{-}11)$$

Note: The IP and SI value estimates for Equations 9-11 through 9-16 should be the same. The differences in these equations are strictly due to rounding errors in conversion values and constants.

To determine the energy demand, use Equation 3-2. For the example,

$$AM_{kW} = K_1 \cdot q_{cap-nom}$$
$$= 0.018 \cdot 762 = 13.7 \, kW$$
$$= 0.005 \cdot 2670 = 13.4 \, kW \quad (9\text{-}12)$$

9.9.2 Cooling Water Pump Electrical Energy Use

Table 9-3 contains information on the size of cooling water pumps required for the given flow. This information is used to estimate the electrical use and demand for the cooling water pump. Equation 3-3, found in chapter 3, was developed to determine the energy use of the cooling water pump. For the example, Equation 3-3 becomes

$$CWP_{kWh} = \frac{K_2 \cdot q_c \cdot H_1 \cdot C_2}{\eta_{CWP}}$$

$$= \frac{0.1 \cdot 800 \cdot 8{,}736 \cdot 0.7457}{0.80} = 651000 \, kWh \quad (9\text{-}13)$$

$$= \frac{0.021 \cdot 2810 \cdot 8736 \cdot 1}{0.80} = 644000 \, kWh$$

Equation 3-2 is used to calculate the electrical demand of the cooling water pump. $K_2 \cdot C_2 / \eta_{CWP}$ is substituted for K_1 in Equation 3-2. For the example,

$$CWP_{kW} = \frac{K_2 \cdot C_2 \cdot q_c}{\eta_{CWP}}$$

$$= \frac{0.1 \cdot 0.7457 \cdot 800}{0.80} = 75 \, kW \quad (9\text{-}14)$$

$$= \frac{0.021 \cdot 1 \cdot 2810}{0.80} = 74 \, kW$$

TABLE 9-3
Cooling Water Pump Sizing

Cooling Water Temperature Difference	Units	Single-Effect Machine	Double-Effect Machine
10°F (6°C)	bhp/ton (kW/kW$_T$)	0.12 (0.025)	0.1 (0.021)
15°F (8°C)	bhp/ton (kW/kW$_T$)	0.09 (0.019)	0.06 (0.013)
20°F (11°C)	bhp/ton (kW/kW$_T$)	0.06 (0.013)	0.05 (0.011)
25°F (14°C)	bhp/ton (kW/kW$_T$)	0.05 (0.011)	0.04 (0.008)
30°F (17°C)	bhp/ton (kW/kW$_T$)	0.04 (0.008)	—
35°F (19°C)	bhp/ton (kW/kW$_T$)	0.04 (0.008)	—

Note: The bhp/ton (kW/kW$_T$) ratings of the cooling water pump(s) are based on a system with a 60-foot head (179 kPa) and a motor that is approximately 80% efficient. Numbers in this table are based on the average of manufacturers' published ratings for cooling water pumps.

APPLICATION GUIDE FOR ABSORPTION COOLING/REFRIGERATION USING RECOVERED HEAT

9.9.3 Cooling Tower Fan Electrical Energy Use

The last large electrical energy use of a LiBr absorption system is the cooling tower. Not all absorption systems will utilize a cooling tower for the heat rejection medium. Instead, the heat in the cooling water could be used to preheat a process in an industrial process or some other similar use. If a cooling tower is used, the energy use and demand for the cooling tower fans must be determined. Table 9-4 contains information to simplify the initial estimating process.

Equation 3-4 from chapter 3 can be used to approximate the cooling fan energy use. Equation 3-4 becomes Equation 9-15 for the example. The variable K_3 is the value from Table 9-4. For the example, a centrifugal unit is chosen for its low noise generation (see Section 3.2.2). Therefore, with a 10°F (6°C) cooling water temperature difference, $K_3 = 0.23$ bhp/ton (0.049 kW/kW$_T$). The variable K_4 is a part-use factor for the cooling tower fans. These data are obtained from Table 3-5 in chapter 3.

With the information from Table 9-4, the cooling tower fan energy use can be approximated using Equation 3-4. For the example, Equation 3-4 becomes

$$\begin{aligned} CTF_{kW} &= \frac{K_3 \cdot q_c \cdot H_1 \cdot C_2 \cdot K_4}{\eta_{CTF}} \\ &= \frac{0.23 \cdot 800 \cdot 8{,}736 \cdot 0.7457 \cdot 0.4}{0.80} = 599000 \, kWh \\ &= \frac{0.049 \cdot 2810 \cdot 8736 \cdot 1 \cdot 0.4}{0.80} = 601000 \, kWh \end{aligned} \quad (9\text{-}15)$$

Equation 3-2 is used to calculate the electrical demand of the cooling tower fan. $C_2 \cdot K_3 / \eta_{CTF}$ is substituted for K_1 in Equation 3-2. For the example,

$$\begin{aligned} CTF_{kW} &= \frac{C_2 \cdot K_3 \cdot q_c}{\eta_{CTF}} \\ &= \frac{0.7457 \cdot 0.23 \cdot 800}{0.80} = 172 \, kW \\ &= \frac{1 \cdot 0.049 \cdot 2810}{0.80} = 172 \, kW \end{aligned} \quad (9\text{-}16)$$

TABLE 9-4
Cooling Tower Fan Sizing

Cooling Water Temperature Difference	Type of Fan	Single-Effect Machine bhp/ton (kW/kW$_T$)	Double-Effect Machine bhp/ton (kW/kW$_T$)
10°F (6°C)	Propeller	0.13 (0.028)	0.11 (0.023)
	Centrifugal	0.27 (0.057)	0.23 (0.049)
15°F (8°C)	Propeller	0.11 (0.023)	0.09 (0.019)
	Centrifugal	0.23 (0.049)	0.19 (0.040)
20°F (11°C)	Propeller	0.09 (0.019)	0.08 (0.017)
	Centrifugal	0.19 (0.040)	0.17 (0.036)

Notes: The bhp/ton (kW/kW$_T$) ratings are per design ton (kW$_T$) of cooling output of the absorption machine and are based on a chilled-water temperature of 44°F (7°C) and the input of 12 psig (184 kPa) steam for a single-effect machine and 115 psig (894 kPa) steam for a double-effect machine. Numbers in this table are based on the average of manufacturers' published ratings for absorption machines and cooling towers.

9.10 CONSIDER OTHER OPTIONS

The example presented in this chapter was based on a double-effect absorption machine with 80°F to 90°F (27°C to 32°C) cooling water, 52°F to 42°F (11°C to 6°C) chilled water, and 115 psig (894 kPa) steam. There are other conditions that could be considered, including different steam pressure, use of hot water instead of steam, different cooling water temperature rise, and use of a single-effect machine.

The results of repeating the application procedures are summarized in Table 9-5. The only difference in the procedures from the previous example is for the hot water alternative. For hot water, Table 9-1 must be used to determine the equivalent steam pressure of the hot water. In the example in Table 9-1, an entering hot water temperature of 360°F (182°C) and a leaving temperature of 320°F (160°C) were used, giving an equivalent steam pressure of 80 psig (653 kPa). See section 9.5 for additional information.

From Table 9-5, several characteristics of absorption machines become apparent. These are as follows.

TABLE 9-5
Comparison of Absorption System Alternatives with a Given Heat Recovery Source

			Base Steam Case	Double-Effect Increased Steam Pressure	Double-Effect Hot Water	Double-Effect Increased Cooling Water ΔT**	Single-Effect Steam
Description	Reference	Units					
1. Determine recovered heat quality	Fig. 9-2 or Tab. 9-1	psig steam (kPa)	115 (894)	130 (998)	80 (635)	115 (894)	12 (184)
2. Calculate amount of recovered heat	Eq. 9-1	Btuh (kW_T)	8,692,000 (2563)	8,571,903 (2528)	9,020,070 (2660)	8,692,454 (2563)	10,065,000*** (2967)
3. Calculate flow of recovered heat	Eq. 9-2	lb/hr (kg/s)	8,094 (1.03)	8,042 (1.02)	8,219 (1.04)	8,094 (1.03)	10,734*** (1.36)
4. Calculate leaving exhaust gas temp.	Eq. 9-3	°F (°C)	390 (199)	398 (203)	369 (187)	390 (199)	300*** (149)
5. Determine cooling output***	Fig. 9-3	tons (kW_T)	800 (2810)	750 (2640)	920 (3236)	800 (2810)	595 (2095)
6. Correction factor for steam pressure	Fig. 9-4 or 5		1	1.08	0.8	1	1
7. Corrected cooling capacity for steam	Eq. 9-4	tons (kW_T)	800 (2810)	694 (2444)	1150 (4045)	800 (2810)	595 (2095)
8. Correction factor for water temps.	Fig. 9-6 or 7		1.05	1.04	1.04	0.79	1.09
9. Corrected rated cooling capacity	Eq. 9-6	tons (kW_T)	762 (2670)	667 (2350)	1,106 (3889)	1,013 (3557)	546 (1920)
10. Flow of cooling water	Tab. 9-2	gpm/ton (L/s·kW_T)	4.5 (0.08)	4.5 (0.08)	4.5 (0.08)	2.3 (0.04)	6.0 (0.1)
11. Amount of heat rejection	Eq. 9-8	Btuh (kW_T)	18,000,000 (5230)	16,875,000 (4890)	20,700,000 (5997)	18,400,000 (5210)	17,900,000 (5246)
12. Energy* factor of absorption machine	Tab. 9-3	kW/ton (kW/kW_T)	0.018 (0.005)	0.018 (0.005)	0.015 (0.004)	0.015 (0.004)	0.012 (0.003)
13. Energy* use/year of absorption machine	Eq. 9-10	kWh/yr	121,000	99,000	145,000	133,000	57,000
14. Energy* demand of absorption machine	Eq. 9-11	kW	14	11	17	15	7
15. Energy* factor of cooling water pump	Tab. 9-2	bhp/ton (kW/kW_T)	0.1 (0.021)	0.1 (0.021)	0.1 (0.021)	0.05 (0.011)	0.12 (0.025)

LiBr Absorption Machine Applications

16. Energy* use/year of cooling water pump	Eq. 9-12	kWh/yr	651,000	611,000	749,000	326,000	581,000
17. Energy* demand of cooling water pump	Eq. 9-13	kW	75	70	86	37	66
18. Energy* factor of cooling tower fans	Table 9-4	bhp/ton (kW/kW_T)	0.23 (0.049)	0.23 (0.049)	0.23 (0.049)	0.17 (0.036)	0.27 (0.057)
19. Energy* use of cooling tower fans	Eq. 9-14	kWh/yr	599,000	562,000	689,000	443,000	523,000
20. Energy* demand of cooling tower fans	Eq. 9-15	kW	172	161	197	127	150
21. Total energy* usage		kWh/yr	1,371,000	1,272,000	1,583,000	902,000	1,161,000
22. Total energy* demand		kW	261	242	300	179	223
23. Energy* demand/actual size		kW/ton (kW/kW_T)	0.326 (0.093)	0.323 (0.092)	0.326 (0.093)	0.224 (0.064)	0.375 (0.107)
24. Absorption machine first cost	Eq. 9-17	$	352,000	315,000	480,000	425,000	193,000
25. Heat rejection equipment first cost	Eq. 9-19	$	176,000	165,000	202,000	152,000	150,000
26. Heat recovery equipment first cost	Eq. 9-21	$	201,000	205,000	202,000	260,000	
27. Total first cost		$	730,000	681,000	779,000	630,000	
				887,000			
28. First cost per actual cooling output	line 27/line 5	$/ton ($/kW_T)	913 (250)	908 (258)	964 (274)	974 (277)	1,013 (288)

* Energy* refers to electrical energy demand or usage.
** The cooling water temperature is 20°F (12°C) as compared to the 10°F (6°C) for the base case.
*** Original calculation of recovered heat resulted in a leaving exhaust gas temperature below 300°F (149°C). The value shown was recalculated assuming a leaving exhaust gas temperature of 300°F (149°C). The SI units are not a direct conversion. They were calculated using the procedure presented in Chapter 8 and in this chapter.
**** The value determined by Figure 9-3 (line 5) is the peak cooling capacity available at the design conditions (manufacturer's rating). The value determined by Equation 9-6 (line 9) is for selecting a unit based on its ARI rated capacity and is used to estimate first cost of the absorption machine.

- If a lower temperature rise (higher flow rate) in the cooling water temperature is used, then the size of the cooling tower decreases and the capacity of the absorption machine decreases. Therefore, to achieve the desired system cooling load, a larger absorption machine may be required to meet capacity.
- The higher the input steam pressure, the greater the cooling output.
- More heat is recovered for a single-effect system than for a double-effect system since the heat is at a lower grade. However, with the lower COP of the single-effect machine, less cooling output is generated.

9.11 NONCONCURRENT LOADS

When considering nonconcurrent loads, several items are important. The primary concern is the time period when the amount of recovered heat does not match the need of the absorption machine. A second concern is the operation of absorption machines at part loads.

9.11.1 Mismatched Loads

For the case of mismatched loads, the base cooling load during the summer for the office building is met with the absorption machine and the variable loads are met with a vapor-compression system. The excess capacity of the absorption machine—200 tons (700 kW$_T$)—is used for process cooling in the industrial plant. During the winter months, when the office cooling load is met through the use of air-side economizers, the cooling output from the absorption machine is used for the process cooling requirements of the industrial plant. Any excess heat from the heat recovery system is used to heat the office complex.

If the only cooling load on the system was the office complex, then during the summer months the excess cooling capacity could be stored in a thermal storage system for use during peak cooling demand periods. Figure 9-8 shows one possible cooling load profile using thermal storage.

9.11.2 Part-Load Conditions

As the cooling load on the absorption system drops, the required heat input also decreases. Figures 9-9 and 9-10 approximate the required heat input for a given cooling load. Figure 9-9 is for single-effect LiBr machines and Figure 9-10 is for double-effect LiBr machines.

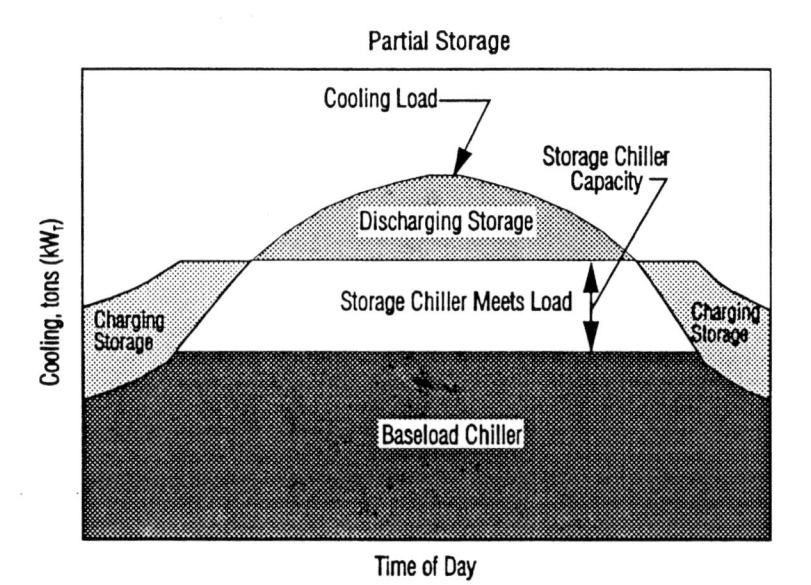

Figure 9-8 Thermal Storage Control Strategy

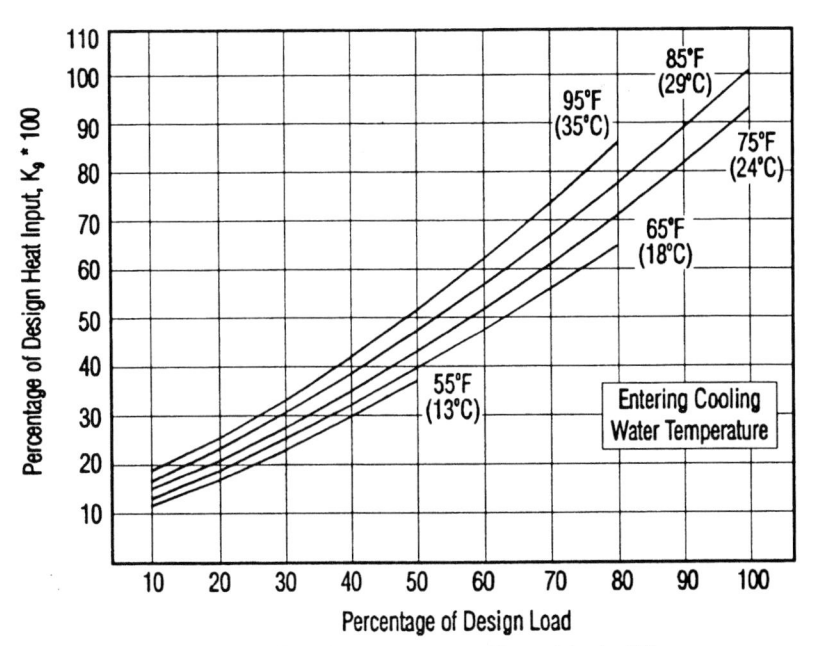

Figure 9-9 Part-Load Capacity Chart, Single-Effect

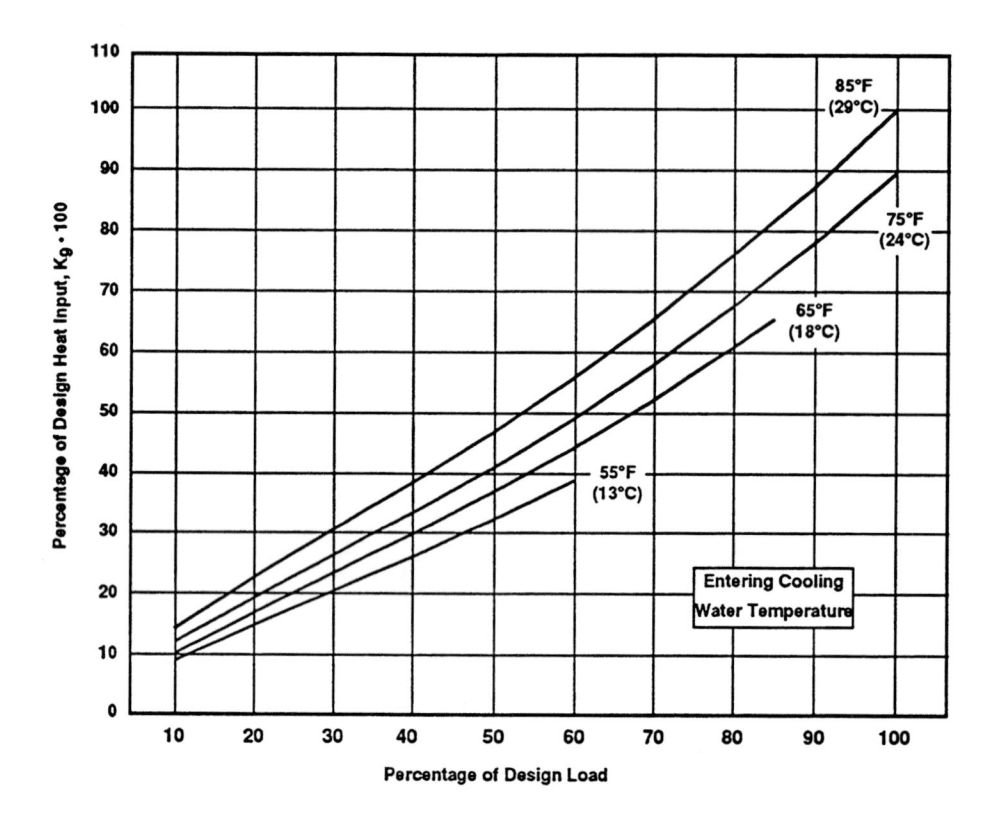

[Figure 9-10 chart: Percentage of Design Heat Input, K₉·100 vs Percentage of Design Load, with curves for Entering Cooling Water Temperature at 85°F (29°C), 75°F (24°C), 65°F (18°C), and 55°F (13°C)]

Figure 9-10 Part-Load Performance, Double-Effect

To calculate the actual heat input, use Equation 9-17:

$$q_{h-part-load} = K_9 \cdot q_{c-part-load} \cdot C_1 \tag{9-17}$$

where

$q_{h\text{-}part\text{-}load}$	=	part-load heat input, Btuh (kW$_T$);
K_9	=	value from Figure 9-9 or 9-10, %;
$q_{c\text{-}part\text{-}load}$	=	part-load cooling output, tons (kW$_T$); and
C_1	=	constant, 12,000 Btuh/ton (1 kW$_T$/kW$_T$).

Using annual load duration curves, the annual estimated fractional use of the available heat can be estimated using Equation 9-17.

9.12 COST ESTIMATES

Using the information presented in chapter 3, the first costs for an absorption system can be estimated. The results from this section are primarily for comparisons of different types of absorption systems.

The primary costs for an absorption system are the absorption machine, heat rejection equipment, heat recovery equipment, and installation and commissioning. The absorption machine first costs can be determined using Equation 9-18 and the information contained in Table 3-1.

$$AM_{COST} = q_{cap-nom} \cdot K_{10} \qquad (9\text{-}18)$$

where

AM_{COST} = absorption machine first cost, \$;
$q_{cap-nom}$ = nominal (ARI) capacity of absorption machine, tons (kW$_T$); and
K_{10} = cost per cooling unit for absorption machine, Table 3-1, \$/ton (\$/kW$_T$).

For the initial example, Equation 9-18 becomes

$$\begin{aligned} AM_{COST} &= 762 \cdot 420 = \$320,000 \\ &= 2670 \cdot 118 = \$315,000 \end{aligned} \qquad (9\text{-}19)$$

Note: Cost estimates for IP and SI units should be the same. Rounding errors in conversion constants result in differences.

The heat rejection equipment first costs can be determined using Equation 9-20. The values for K_{11} are obtained from Table 3-2 in chapter 3.

$$CT_{COST} = q_c \cdot K_{11} \qquad (9\text{-}20)$$

where

CT_{COST} = cooling tower first cost, \$;
q_c = design capacity of absorption machine, tons (kW$_T$)
K_{11} = cost per cooling unit for the cooling tower from Table 3-2, \$/ton (\$kW$_T$).

For the example, Equation 9-20 becomes

$$CT_{COST} = 800 \cdot 220 = \$176,000$$
$$= 2810 \cdot 63 = \$177,000 \qquad (9\text{-}21)$$

The heat recovery equipment costs are highly variable. For initial estimates, use a value of $25/lb/h ($55/kg/s) of steam recovered (from section 3.2.3). Equation 9-22 can be used to estimate the first cost of the heat recovery equipment:

$$HR_{COST} = m_s \cdot K_{12} \qquad (9\text{-}22)$$

where

HR_{COST} = heat recovery equipment cost, $;
m_s = steam flow rate, from Equation 9-2, lb/h (kg/s); and
K_{12} = heat recovery equipment cost per recovered heat flow rate, $ per lb/h ($ per kg/s).

For the example, Equation 9-22 becomes

$$HR_{COST} = 8094 \cdot 25 = \$202,000$$
$$= 1.03 \cdot 198000 = \$204,000 \qquad (9\text{-}23)$$

The installation and commissioning costs for an absorption system will vary depending on site-specific conditions and have not been estimated in this guide.

REFERENCES

Manufacturer. 1994. *U.S. Turbines catalog*.
Plzak, W. 1994. Personal communication (phone).

CHAPTER 10
AAR Machine Applications

10.1 INTRODUCTION

AAR machines are custom designed and constructed for each application. For this reason, there are limited data on the actual sizing and application of these systems. The procedure developed in chapter 8 can be used on AAR machines. However, the level of detail and the number of variations accounted for in LiBr systems in chapter 9 are not attainable.

Therefore, this chapter has a simplified application procedure based on the maximum COP attainable for a given set of conditions. Appendix B contains a complete discussion of the algorithms used to produce the charts used in this chapter. An example has been included to show the application of the procedure developed in chapter 8 for AAR machines.

10.2 EXAMPLE DESCRIPTION

The example presented is for a cogeneration plant located in central Texas. A gas turbine is used to generate electricity, with excess electricity retailed to the local utility. Recovered heat is obtained from the exhaust stack of the gas turbine. The output from the AAR machine is 4°F (–16°C) ammonia, which is used to cool a nearby food cold storage warehouse. The heat rejection device for the absorption machine is an evaporative cooling tower. The system layout is shown in Figure 10-1.

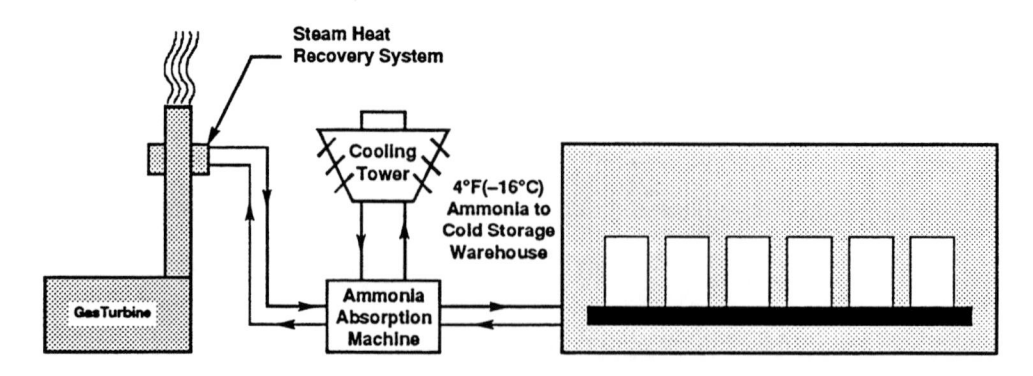

Figure 10-1 Layout for AAR Example

For the central Texas location, the outdoor design wet-bulb temperature is 73°F (23°C). With a 7°F (4°C) approach on the evaporative condenser, this equates to an entering condenser water temperature to the absorption machine of 80°F (27°C).

The gas turbine produces approximately 2.3 MW at ambient conditions. At the ambient outdoor temperature of 95°F (35°C), the exhaust gas of the gas turbine is approximately 1,055°F (570°C) at 106,000 pounds per hour (13.4 kg/s) (Manufacturer 1994a).

10.3 CALCULATE RECOVERABLE HEAT

Before determining the amount of heat that can be recovered, the optimum temperature of the generator must first be determined. To calculate the optimum temperature of the generator, the temperature of the ammonia condensate leaving the condenser and the evaporator temperature must be determined. The temperature of the ammonia condensate can be estimated to be 5°F (3°C) above the temperature of the entering condenser water. This gives an ammonia condensate temperature for the example of 85°F (29°C).

The charts in Figures 10-2 and 10-3 have been developed from unpublished research by Vicatos et al. (1994). The starting point is the intersection of the ammonia condensate temperature (T_s) curve and the evaporator temperature (T_e) curve. From this intersection, lines are drawn parallel to the x- and y-axes until the lines intersect with the ammonia condensate temperature on the z-axis. These two intersection points have lines drawn vertically down to the x- and y-axes to determine the optimum generator temperature and the maximum COP that corresponds to the generator temperature. For the example,

T_s = 85°F (25°C) and
T_e = 4°F (–15°C).

The resultant optimum generator temperature and corresponding maximum COP are

T_g = 198°F (92°C) and
COP_{max} = 0.59.

    The estimated generator temperature is the temperature of the strong solution (weak aqua) leaving the generator. The actual temperature of the steam entering the generator can be estimated to be 25°F (14°C) higher than the generator temperature. Therefore, for the example, the entering (saturated) steam temperature is approximately 223°F (106°C).

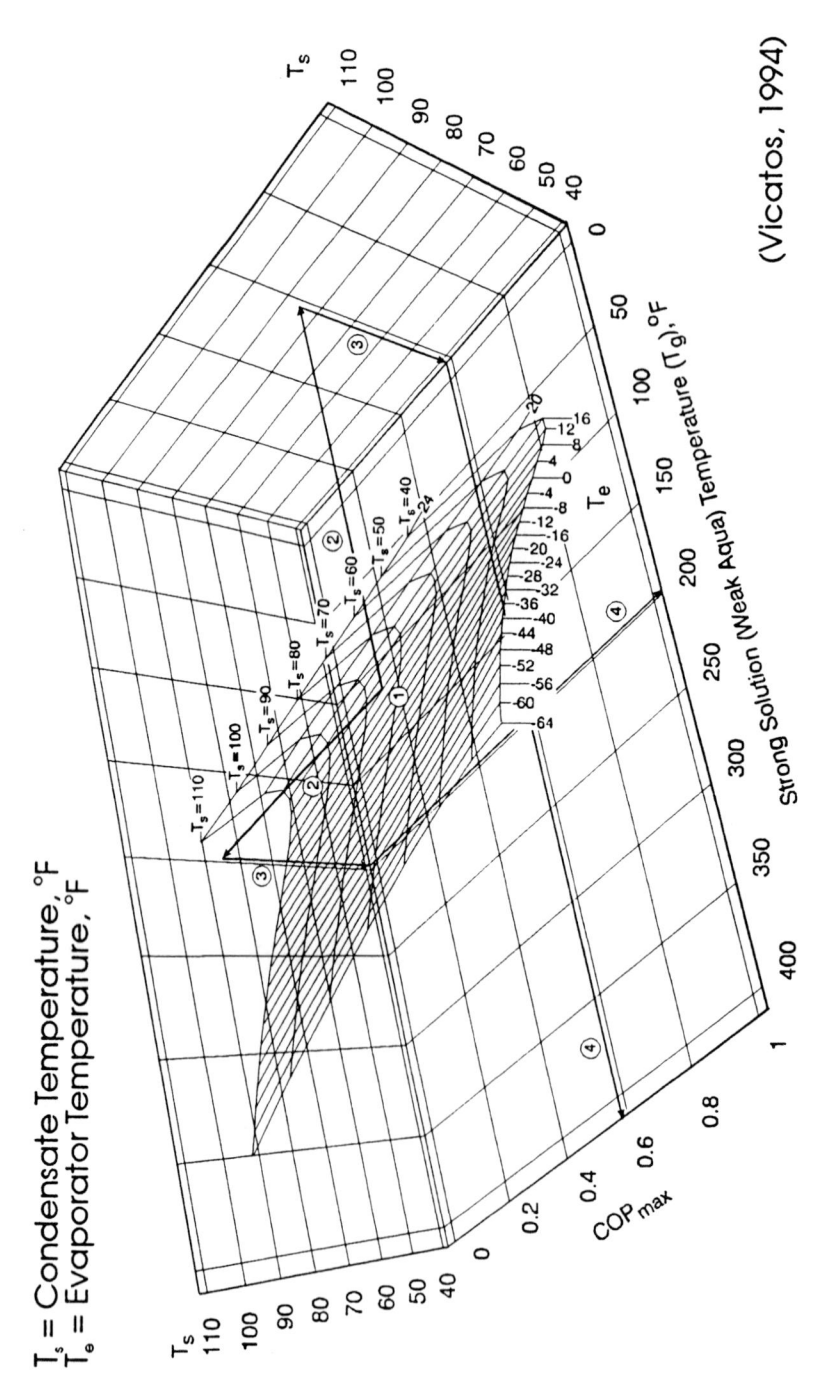

Figure 10-2 AAR Optimization Chart, IP Units

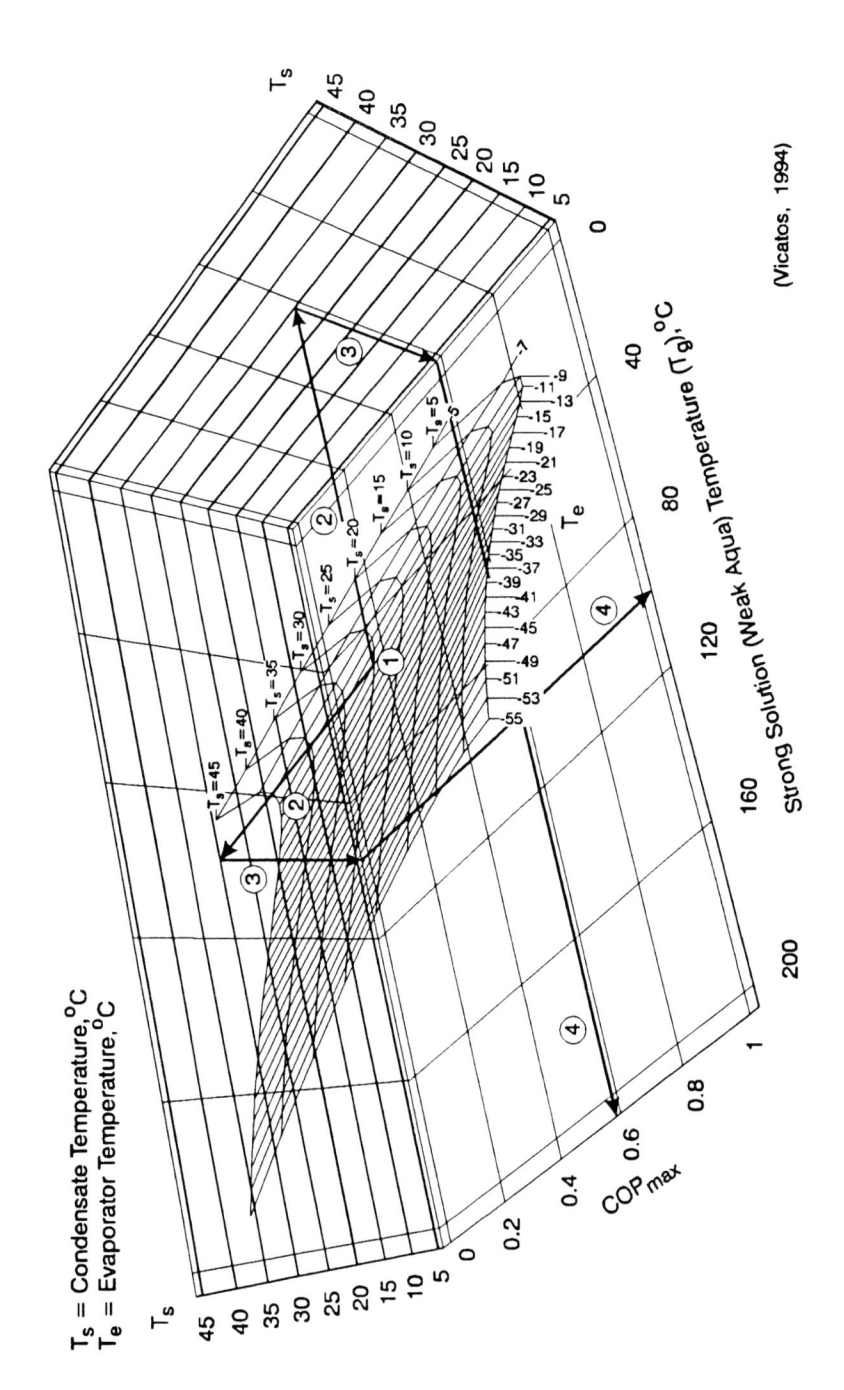

Figure 10-3 AAR Optimization Chart, SI Units

135

Since the heat recovery system is a gas-to-gas (steam) system, Equation 8-1 can be used to determine the quantity of heat recovered. Therefore, Equation 8-1, found in chapter 8, becomes

$$\begin{aligned} q &= \eta_{eff}(c_{p,gas})(m_{gas})\left(T_{in,gas} - T_{mean,steam}\right) \\ &= 0.93\,(0.25)\,(106{,}000)\,(1{,}055 - 223) = 20{,}504{,}600\,Btuh \\ &= 0.93\,(1.05)\,(13.4)\,(570 - 106) = 6070\,kW_T \end{aligned} \quad (10\text{-}1)$$

To ensure that condensation does not occur in the exhaust gas stack of the gas turbine, the exhaust gas temperature must be kept above 300°F (149°C). To calculate the leaving exhaust gas temperature from the heat recovery system, use Equation 8-2. Equation 8-2 becomes

$$\begin{aligned} T_{gas,leaving} &= T_{gas,entering} - \frac{q_{recovered}}{c_{p,gas}m_{gas}} \\ &= 1{,}055 - \frac{20{,}504{,}600}{0.25\,(106{,}000)} = 281°F \\ &= 570 - \frac{6070}{1.05\,(13.4)} = 139°C \end{aligned} \quad (10\text{-}2)$$

Therefore, since the leaving exhaust gas temperature is less than 300°F (149°C), the amount of recoverable heat must be recalculated using a leaving exhaust gas temperature of 300°F (149°C). The results are shown in Equation 10-3, which assumes that the effectiveness of the heat recovery system is lower than that used to calculate the amount of recoverable heat. This assumption ensures that the other values in Equation 10-2 remain constant (temperature of the entering gas and mass flow rate of the gas).

$$\begin{aligned} q_{recovered} &= \left(T_{gas,entering} - T_{gas,leaving}\right)(c_{p,gas})(m_{gas}) \\ &= (1{,}055 - 300)\,(0.25)\,(106{,}000) = 20{,}007{,}500\,Btuh \\ &= (570 - 149)\,(1.05)\,(13.4) = 5925\,kW_T \end{aligned} \quad (10\text{-}3)$$

10.4 DETERMINE AMOUNT OF COOLING OUTPUT

Before determining the amount of cooling output by using Equation 10-4, the effect of superheating and subcooling of the steam must be determined. Superheating of the steam ensures that saturated steam is supplied to the absorption machine. Subcooling of the condensate is required so the condensate does not flash to steam in the return lines.

A correction factor, K_5, has been included in Equation 10-4 to account for the effect of subcooling and superheating of the steam. For initial estimates, a value of 1.25 can be used for K_5.

$$q_{cooling} = \frac{(q_{recovered})(COP_{max})}{(C_1)(K_5)} \tag{10-4}$$

where

$q_{cooling}$	=	cooling output of AAR machine, tons (kW_T);
$q_{recovered}$	=	heat recovered, Btuh (kW_T);
COP_{max}	=	maximum COP value from Figure 10-2 or 10-3;
C_1	=	conversion from Btuh to tons, 12,000 Btuh/ton (1 kW_T/kW_T); and
K_5	=	correction factor for superheating and subcooling, %.

For the example,

$$q_{cooling} = \frac{(20{,}007{,}500)(0.59)}{(12{,}000)(1.25)} = 787 \; tons$$

$$= \frac{(5925)(0.59)}{(1.0)(1.25)} = 2797 \; kW_T \tag{10-5}$$

10.5 OTHER SIZING PARAMETERS

Since all AAR systems are customed designed and constructed for each application, further detailed information for the application of AAR systems is not available since manufacturer selection catalogs are not available in the form in which average performance data could be developed. Manufacturers must be consulted to determine the heat rejection and electric power requirements, part-load and nonconcurrent load operation, and the use of recovered heat in the form of a liquid and not a gas.

For rough approximations of the additional requirements of AAR systems, Table 10-1 has been included. This table is based on the available manufacturers' published literature (Manufacturer 1994b).

REFERENCES

Manufacturer. 1994a. *U.S. Turbines catalog*.
Manufacturer. 1994b. North Salt Lake City, UT: Lewis Energy Systems, Inc.
Vicatos, G., and J. Gryzagoridis. 1994. Unpublished research. Department of Mechanical Engineering. Cape Town, South Africa: University of Cape Town.

TABLE 10-1
AAR Machine Characteristics

	\multicolumn{11}{c}{Evaporator Temperature, °F (°C)}
	50 (10)	40 (4)	30 (-1)	20 (-7)	10 (-12)	0 (-18)	-10 (-23)	-20 (-29)	-30 (-34)	-40 (-40)	-50 (-46)
One-Stage											
Steam Pressure, psia (kPa)	14 (97)	20 (138)	24 (165)	31 (214)	42 (290)	53 (365)	67 (462)	83 (572)	103 (710)	135 (931)	173 (1193)
Steam Sat. Temp., °F (°C)	210 (99)	225 (107)	240 (116)	255 (124)	270 (132)	285 (141)	300 (149)	315 (157)	330 (166)	350 (177)	370 (188)
Generator Heat Required, Btu/min·ton (kJ/s·kW$_T$)	300 (1.5)	325 (1.6)	347 (1.7)	373 (1.9)	400 (2.0)	430 (2.2)	466 (2.3)	511 (2.6)	571 (2.9)	645 (3.2)	754 (3.8)
Steam Rate, lb/h·ton (g/s·kW$_T$)	19 (0.68)	20 (0.71)	22 (0.79)	24 (0.86)	26 (0.93)	28 (1.00)	31 (1.11)	34 (1.22)	39 (1.40)	45 (1.61)	53 (1.90)
Condenser, gpm/ton 87–105°F (mL/s·kW$_T$ 31–41°C)	3.6 (65)	3.7 (66)	3.8 (68)	4.0 (72)	4.3 (77)	4.5 (81)	5.0 (90)	5.5 (99)	6.1 (109)	7.1 (127)	8.8 (158)
Two-Stage											
Steam Pressure, psia (kPa)	6.7 (46)	7.5 (52)	9.3 (64)	10 (69)	12 (83)	14 (97)	17 (117)	20 (138)	25 (172)	30 (207)	39 (269)
Steam Sat. Temp., °F (°C)	175 (79)	180 (82)	190 (88)	195 (91)	205 (96)	210 (99)	220 (104)	230 (110)	240 (116)	250 (121)	265 (129)
Generator Heat Required, Btu/min·ton (kJ/s·kW$_T$)	550 (2.8)	577 (2.9)	605 (3.0)	637 (3.2)	670 (3.4)	711 (3.6)	753 (3.8)	799 (4.0)	850 (4.3)	905 (4.5)	970 (4.9)
Steam Rate, lb/h·ton (g/s·kW$_T$)	33 (1.2)	35 (1.3)	37 (1.3)	39 (1.4)	41 (1.5)	44 (1.6)	47 (1.7)	50 (1.8)	54 (1.9)	58 (2.1)	62 (2.2)
Condenser, gpm/ton 87–105°F (mL/s·kW$_T$ 31–41°C)	4.0 (72)	4.2 (75)	4.3 (77)	4.5 (81)	4.9 (88)	5.3 (95)	5.8 (104)	6.4 (115)	7.2 (129)	8.3 (149)	10.2 (183)

CHAPTER 11
Other Technologies

Absorption technology may be used in many different ways. This chapter discusses some less well-known applications and new developments in absorption technology. The information in this chapter is included to provide a listing of known absorption technologies and to prevent them from not being investigated when they may be applicable.

11.1 TRIPLE EFFECT

Triple-effect machines are currently being developed by manufacturers and researchers. However, these units are currently not commercially available. The description in this chapter is for one of the systems currently under development. Discussion of other system configurations can be found in other published reports (ASME 1994).

In a triple-effect cycle, higher COPs are obtained by adding a topping cycle to a double-effect LiBr machine, a third effect, or another cycle combination. Due to the high temperatures required by a triple-effect machine, a fluid different than LiBr is used in the topping cycle. One such fluid that has received much attention is Alkitrate, a water and ternary blend of nitrate salts ($LiNO_3$, KNO_3, and $NaNO_3$), patented in 1986.

In order to achieve three effects (use the heat three times), the heat of condensation from the topping-cycle refrigerant and the heat produced in the topping cycle absorber section are used to power the high-stage generator of the double-effect cycle. The heat of condensation for the high-stage refrigerant is used to power the low-stage generator, just as in a double-effect LiBr machine.

The refrigerant for the system is shared by all three parts of a triple-effect machine (topping, high stage, and low stage). Due to the high temperatures needed to power the topping cycles, the triple-effect systems currently under development will all be direct-fired machines. A typical generator temperature of approximately 445°F (250°C) is used for the heat input to the topping cycle. The attainable cooling COP for a triple-effect machine is approximately 1.5.

Currently, triple-effect machines are in the experimental stages. Manufacturers estimate that commercial units will be available in 1996 (GRI 1990; Ouimette and Herold 1994; Plzak 1994; Ryan 1994; Reimann 1994).

11.2 GAX CYCLE

In recent years, generator-absorber heat exchanger (GAX) cycles have had increasing attention. A GAX cycle uses absorber cooling water to heat the generator. A primary benefit of a GAX cycle is that its performance varies continuously with temperature lift, which results in a high seasonal performance. The COP for a GAX cycle can be 70% to 130% greater than that of a single-effect machine (Rane and Erickson 1994).

The increase in the COP is obtained by utilizing some of the heat generated in the absorber section to heat the fluid in the generator section. This reduces the amount of heat input required and the cooling tower load while maintaining the level of cooling output (Inoue et al. 1994).

11.3 HEATING

Absorption machines can be used as heat pumps to increase waste heat temperature. Heat from the absorber is added to the waste heat and this combined heat is provided to the process or load. Two applications for increased heat include process/boiler feed water heating and evaporation/concentration of liquids or distillation.

Absorption heat pumps often use LiBr-water solutions, but many other working fluid pairs have been used in installations and for research and development purposes. Two of the more popular fluid pairs for absorption heat pumps are LiBr/H_2O and NaOH/KOH/H_2O. In addition, research has been completed on a working-fluid pair of TFE (2,2,2-trifluor-ethanol) as the refrigerant and Pyr (2-pyrrolidone) as the absorbent (Machielsen 1990). Also, the working pairs of ammonia/lithium nitrate and NaSCN/NH_3 (sodium thiocyanate/ammonia) are being investigated (Best et al. 1990; Manufacturer 1993). An absorption heat pump can be either a heat-amplifying (HA) or temperature-amplifying (TA) machine.

11.3.1 Heat Amplifier

Occasionally referred to as a forward absorption heat pump (AHP) or type I AHP, a heat amplifier combines high- and low-temperature heat and delivers medium-temperature heat. Figure 11-1 (RCG/Hagler 1990) illustrates this process. Heat amplifiers typically produce temperatures between 100°F and 302°F (38°C and 150°C).

11.3.2 Temperature Amplifier

The temperature amplifier, sometimes referred to as a heat transformer, temperature booster, or type II or reverse AHP, accepts medium-temperature heat (usually waste) and delivers heat at both higher and lower temperatures. Figure 11-2 shows a temperature amplifier (Machielsen 1990).

OTHER TECHNOLOGIES

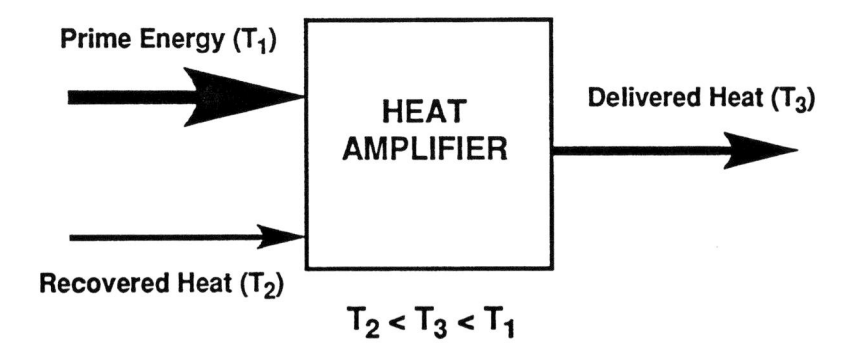

Figure 11-1 Heat Amplifier

$T_2 < T_3 < T_1$

Approximately 40% of the waste heat supplied to a temperature amplifier can be upgraded in temperature by about 100°F (55°C).

In research conducted by Ikeuchi et al. (1985), waste heat at 140°F to 194°F (60°C to 90°C) was boosted in temperature to as high as 302°F (150°C). This was accomplished in part through the use of specially wound helical segmented fins in the heat exchanger.

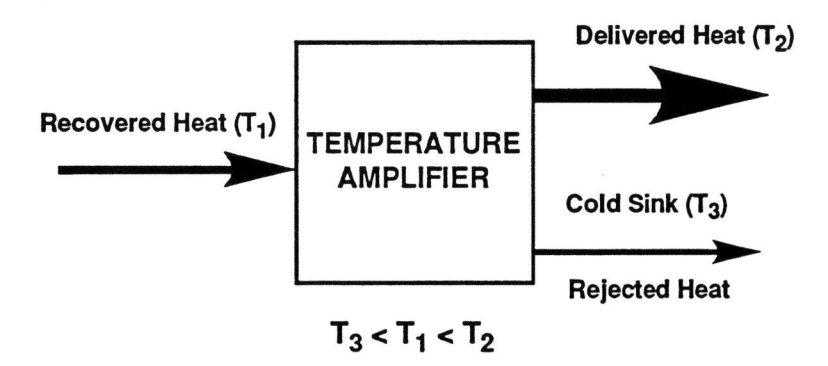

Figure 11-2 Temperature Amplifier

$T_3 < T_1 < T_2$

In other research, process heat was provided by an absorption heat transformer using water preheated in solar ponds. The water temperature at the pond bottom was 185°F (85°C). The temperature boost was to approximately 248°F (120°C) with a COP of approximately 0.5 using a single-effect LiBr/water absorber. A double-effect LiBr absorption machine boosted the temperature to 302°F (150°C) at a COP of approximately 0.3. At the top of the pond, cooler (86°F [30°C]) water was used for the condenser (Grossman 1991).

Some applications of AHPs in the paper industry produce one unit of 240°F (115°C) steam from 10 units of 140°F (60°C) heat stream. These were specially built units constructed in the 1980s and have not received additional applications.

11.4 COMBINATION HEATING AND COOLING

In certain circumstances, the absorption process using waste heat may be able to provide high-temperature fluid to one process and chilled water to another. Similarly, an absorption unit may be used to provide heat in the winter and cooling in the summer. These custom dual-use systems can be cost effective. However, first costs are high and maintenance has been high on the limited number of units manufactured. Even though these applications have been unique, the technology should be investigated if the situation warrants.

REFERENCES

ASME. 1994. *Proceedings of the International Absorption Heat Pump Conference*, AES vol. 31. New York: American Society of Mechanical Engineers.

Best, R., W. Rivera, I. Pilatowsky, and F. A. Holland. 1990. Thermodynamic design data for absorption heat transformers—Part four: Operating on ammonia-lithium nitrate. *Heat Recovery Systems & CHP* 10(5/6): 539–548.

GRI. 1990. Triple-effect absorption chiller. Tech Profile 1290LB12000. Chicago: Gas Research Institute.

Grossman, G. 1991. Absorption heat transformers for process heat generation from solar ponds. *ASHRAE Transactions* 97(1): 420–427.

Ikeuchi, N., T. Yumikura, E. Ozaki, and G. Yamanaka. 1985. Design and performance of a high-temperature-boost absorption heat pump. *ASHRAE Transactions* 91(2B): 2081–2094.

Inoue, N., H. Iizuka, Y. Nimomiya, K.-I. Wantabe, and T. Acki. 1994. *COP evaluation for advanced ammonia-based absorption cycles*. AES #31, pp. 1–6.

Machielsen, C.H.M. 1990. Research activities on absorption systems for heating, cooling and industrial use. *ASHRAE Transactions* 96(1): 1577–1581.

Manufacturer. 1993. *Double-effect absorption gas heat pump—Residential and commercial prospectus*. Columbus, OH: Columbia Gas Distribution Companies.

Ouimette, M. S., and K. E. Herold. 1994. *Performance modeling of triple-effect absorption chiller*. AES #31, pp. 233–241.

Plzak, W. 1994. Personal communication (phone).

Rane, M. V., and D. C. Erickson. 1994. *Advanced absorption cycle: Vapor exchange GAX*. AES #31, pp. 25–32.

RCG/Hagler, Bailly, Inc. 1990. Opportunities for industrial chemical heat pumps in process industries. Prepared for the U.S. Department of Energy, contract no. DE-AC01-87CE40762, subcontract no. 3090-1.

Reimann, B. 1994. Personal communication (letter).

Ryan, W. 1994. Personal communication (letter).

APPENDIX A
Property Charts, Tables, and Conversions

TABLE A-1
Units and Conversions

Dimension	IP Unit	Equals	SI Unit
Temperature	°F	(°F – 32)/1.8	°C
Temperature Difference	Δ°F	/1.8	ΔK (Δ°C)
Volume Flow	gpm	* 0.0631	L/s
Energy Transfer Rate	Btu/h	* 0.2931 – 10^{-3}	kW_T
	Btu/h	* 0.2931 – 10^{-6}	MW_T
Pressure	psia	* 6.8946	kPa
	psig	(psig + 14.696) * 6.895	kPa
Specific Heat	Btu/lb•°F	* 4.184	kJ/(kg•K)
Mass Flow Rate	lb/h	0.126 – 10^{-3}	kg/s
Cooling Capacity	Btuh	/12,000	tons
	ton	3.517	kW_T
Enthalpy	Btu/lb	* 2.326	kW_T•s/kg
Refrigeration	gpm/ton	* 0.0179	L/s•kW_T
Heat to Cooling	Btu/min/ton	* 0.005	kJ/s•kW_T
Steam Rate	lb/h/ton	* 0.03583	g/s•kW_T
	Btuh/kW	* 0.2931 – 10^{-3}	kJ/s•kW
Motor Use	bhp/ton	* 0.212	kW/kW_T
Cost Factor	$/ton	* 0.284	$/$kW_T$

TABLE A-2
Steam Tables and Water Refrigerant Properties—Temperature Arrangement

Steam Temperature, °F (°C)	Steam Pressure, psia (kPa)	Condensate Enthalpy, h_f, Btu/lb (kW•s/kg)	Steam Enthalpy, h_g, Btu/lb (kW•s/kg)
40 (4)	0.122 (0.841)	8.0 (18.6)	1,078.5 (2508.6)
50 (10)	0.178 (1.227)	18.1 (42.1)	1,082.9 (2518.8)
60 (16)	0.256 (1.765)	28.0 (65.1)	1,087.3 (2529.1)
70 (21)	0.363 (2.503)	38.0 (88.4)	1,091.7 (2539.3)
80 (27)	0.507 (3.496)	48.0 (111.6)	1,096.0 (2549.3)
90 (32)	0.699 (4.820)	58.0 (134.9)	1,100.3 (2559.3)
100 (38)	0.650 (4.482)	68.0 (158.2)	1,104.6 (2569.3)
110 (43)	1.276 (8.798)	78.0 (181.4)	1,108.9 (2579.3)
120 (49)	1.694 (11.680)	88.0 (204.7)	1,113.2 (2589.3)
130 (54)	2.224 (15.334)	98.0 (227.9)	1,117.4 (2599.1)
140 (60)	2.891 (19.933)	108.0 (251.2)	1,121.6 (2608.8)
150 (66)	3.720 (25.649)	118.0 (274.5)	1,125.7 (2618.4)
160 (71)	4.743 (32.703)	128.0 (297.7)	1,129.8 (2627.9)
170 (77)	5.994 (41.329)	138.0 (321.0)	1,133.9 (2637.5)
180 (82)	7.513 (51.802)	148.0 (344.2)	1,137.9 (2646.8)
190 (88)	9.341 (64.406)	158.0 (367.5)	1,141.9 (2656.1)
200 (93)	11.526 (79.472)	168.1 (391.0)	1,145.8 (2665.1)
210 (99)	14.122 (97.371)	178.2 (414.5)	1,149.6 (2674.0)
220 (104)	17.184 (118.48)	188.2 (437.8)	1,153.4 (2682.8)
230 (110)	20.774 (143.24)	198.3 (461.2)	1,157.0 (2691.2)
240 (116)	24.960 (172.10)	208.5 (485.0)	1,160.6 (2699.6)
250 (121)	29.814 (205.57)	218.6 (508.5)	1,164.1 (2707.7)
260 (127)	35.411 (244.16)	228.8 (532.2)	1,167.6 (2715.8)
270 (132)	41.835 (288.45)	239.0 (555.9)	1,170.9 (2723.5)
280 (138)	49.173 (339.05)	249.2 (579.6)	1,174.1 (2731.0)
290 (143)	57.516 (396.57)	259.5 (603.6)	1,177.2 (2738.2)
300 (149)	66.963 (461.71)	269.8 (627.6)	1,180.2 (2745.1)
310 (154)	77.615 (535.16)	280.1 (651.5)	1,183.0 (2751.7)
320 (160)	89.580 (617.65)	290.5 (675.7)	1,185.7 (2757.9)
330 (166)	102.97 (709.98)	300.9 (699.9)	1,188.3 (2764.0)
340 (171)	117.90 (812.92)	311.3 (724.1)	1,190.7 (2769.6)
350 (177)	134.50 (927.38)	321.8 (748.5)	1,193.0 (2774.9)
360 (182)	152.89 (1054.178)	332.4 (773.2)	1,195.1 (2779.8)
370 (188)	173.20 (1194.21)	342.9 (797.6)	1,197.0 (2784.2)
380 (193)	195.57 (1348.46)	353.6 (822.5)	1,198.7 (2788.2)
390 (199)	220.14 (1517.87)	364.3 (847.4)	1,200.3 (2791.9)
400 (204)	247.05 (1703.41)	375.1 (872.5)	1,201.6 (2794.9)

TABLE A-3
Steam Tables and Water Refrigerant Properties—Pressure Arrangement

Steam Pressure, psia (kPa)	Steam Temperature, °F (°C)	Condensate Enthalpy, h_f, Btu/lb (kW·s/kg)	Steam Enthalph, h_g Btu/lb (kW·s/kg)
1 (108)	101.7 (38.7)	69.7 (162.1)	1,106.0 (2571.9)
2 (115)	126.1 (52.3)	94.0 (218.6)	1,116.2 (2595.6)
3 (122)	141.5 (60.8)	109.4 (254.3)	1,122.6 (2610.5)
4 (129)	153.0 (67.2)	120.9 (281.0)	1,127.3 (2621.4)
5 (136)	162.2 (72.4)	130.1 (302.6)	1,131.1 (2630.3)
6 (143)	170.1 (76.7)	138.0 (320.8)	1,134.2 (2637.5)
7 (150)	176.9 (80.5)	144.8 (336.6)	1,136.9 (2643.7)
8 (156)	182.9 (83.8)	150.8 (350.6)	1,139.3 (2649.3)
9 (163)	188.3 (86.8)	156.2 (363.3)	1,141.4 (2654.2)
10 (170)	193.2 (89.6)	161.2 (374.8)	1,143.3 (2658.6)
14.696 (202.646)	212.0 (100.0)	180.1 (418.7)	1,150.4 (2675.1)
15 (205)	213.0 (100.6)	181.1 (421.2)	1,150.8 (2676.1)
20 (239)	228.0 (108.9)	196.2 (456.2)	1,156.3 (2688.9)
25 (274)	240.1 (115.6)	208.4 (484.7)	1,160.6 (2698.9)
30 (308)	250.3 (121.3)	218.8 (508.8)	1,164.1 (2707.0)
35 (343)	259.3 (126.3)	227.9 (530.0)	1,167.1 (2714.0)
40 (377)	267.3 (130.7)	236.0 (548.9)	1,169.7 (2720.0)
45 (412)	274.4 (134.7)	243.4 (565.9)	1,172.0 (2725.4)
50 (446)	281.0 (138.3)	250.1 (581.6)	1,174.1 (2730.2)
55 (481)	287.1 (141.7)	256.3 (596.0)	1,175.9 (2734.4)
60 (515)	292.7 (144.8)	262.1 (609.5)	1,177.6 (2738.4)
65 (549)	298.0 (147.8)	267.5 (622.0)	1,179.1 (2741.9)
70 (584)	302.9 (150.5)	272.6 (633.9)	1,180.6 (2745.4)
75 (618)	307.6 (153.1)	277.4 (645.1)	1,181.9 (2748.4)
80 (653)	312.0 (155.6)	282.0 (655.8)	1,183.1 (2751.2)
85 (687)	316.3 (157.9)	286.4 (666.0)	1,184.2 (2753.7)
90 (722)	320.3 (160.2)	290.6 (675.7)	1,185.3 (2756.3)
95 (756)	324.1 (162.3)	294.6 (685.0)	1,186.2 (2758.4)
100 (791)	327.8 (164.3)	298.4 (693.9)	1,187.2 (2760.7)
110 (860)	334.8 (168.2)	305.7 (710.8)	1,188.9 (2764.7)
120 (929)	341.3 (171.8)	312.4 (726.5)	1,190.4 (2768.2)
130 (998)	347.3 (175.2)	318.8 (741.4)	1,191.7 (2771.2)
140 (1067)	353.0 (178.3)	324.8 (755.3)	1,193.0 (2774.2)
150 (1136)	358.4 (181.3)	330.5 (768.6)	1,194.1 (2776.8)
160 (1204)	363.5 (184.2)	335.9 (781.2)	1,195.1 (2779.1)
170 (1273)	368.4 (186.9)	341.1 (793.2)	1,196.0 (2781.2)
180 (1342)	373.1 (189.5)	346.0 (804.7)	1,196.9 (2783.3)

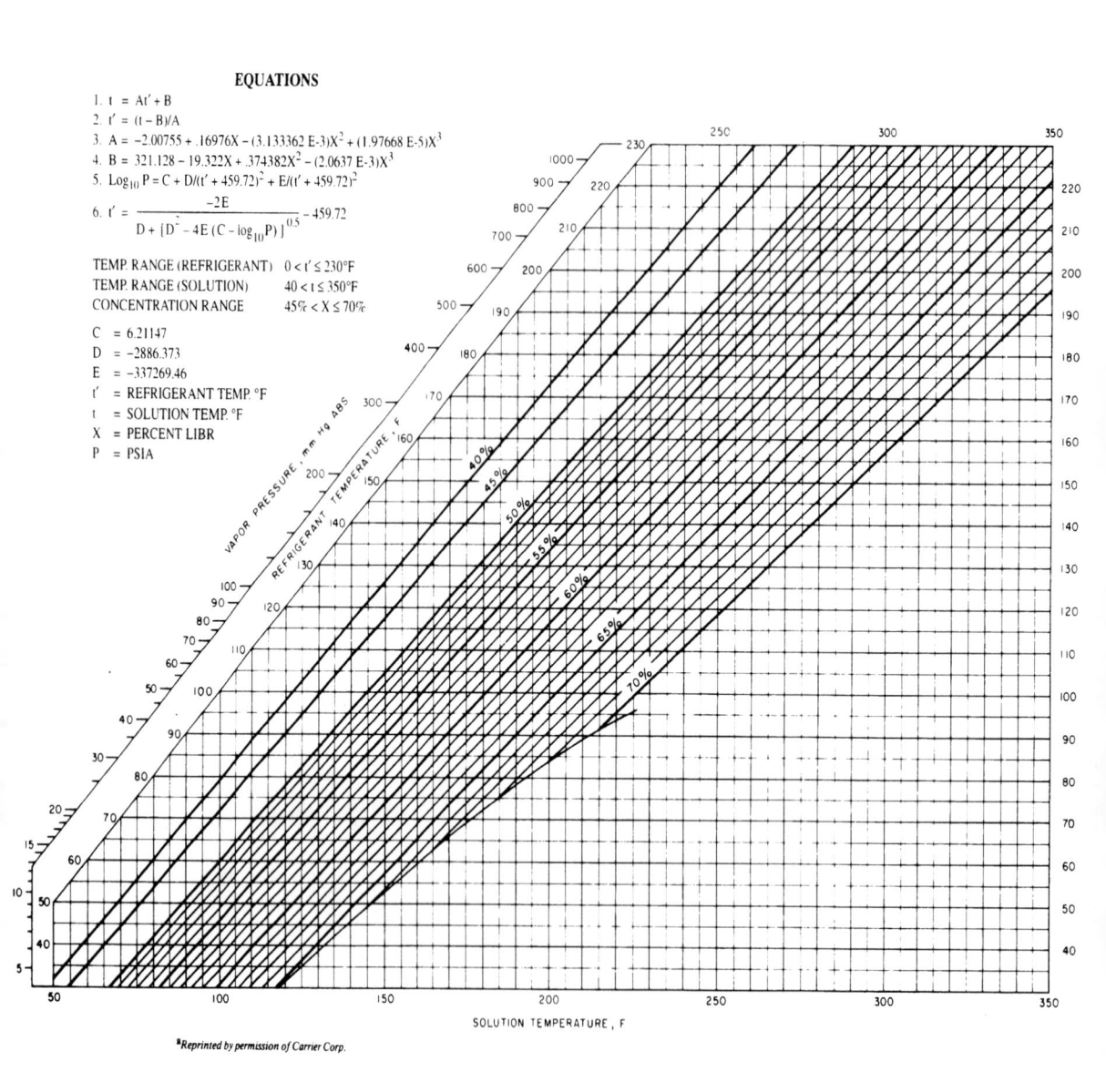

APPLICATION GUIDE FOR ABSORPTION COOLING/REFRIGERATION USING RECOVERED HEAT

EQUATIONS

1. $t = At' + B$
2. $t' = (t - B)/A$
3. $A = -2.00755 + .16976X - (3.133362\ E\text{-}3)X^2 + (1.97668\ E\text{-}5)X^3$
4. $B = 321.128 - 19.322X + .374382X^2 - (2.0637\ E\text{-}3)X^3$
5. $\log_{10} P = C + D/(t' + 459.72)^2 + E/(t' + 459.72)^2$
6. $t' = \dfrac{-2E}{D + [D^2 - 4E(C - \log_{10}P)]^{0.5}} - 459.72$

TEMP. RANGE (REFRIGERANT) $0 < t' \leq 230°F$
TEMP. RANGE (SOLUTION) $40 < t \leq 350°F$
CONCENTRATION RANGE $45\% < X \leq 70\%$

$C = 6.21147$
$D = -2886.373$
$E = -337269.46$
t' = REFRIGERANT TEMP. °F
t = SOLUTION TEMP. °F
X = PERCENT LIBR
P = PSIA

[a]Reprinted by permission of Carrier Corp.

Figure A-1 Equilibrium Chart for Aqueous Lithium Bromide Solutions
Source: ASHRAE. 1993. *Handbook of Fundamentals*. American Society of Heating, Refrigerating and Air-Conditioning Engineers, Inc., Atlanta, GA, Chapter 17, Figure 40)

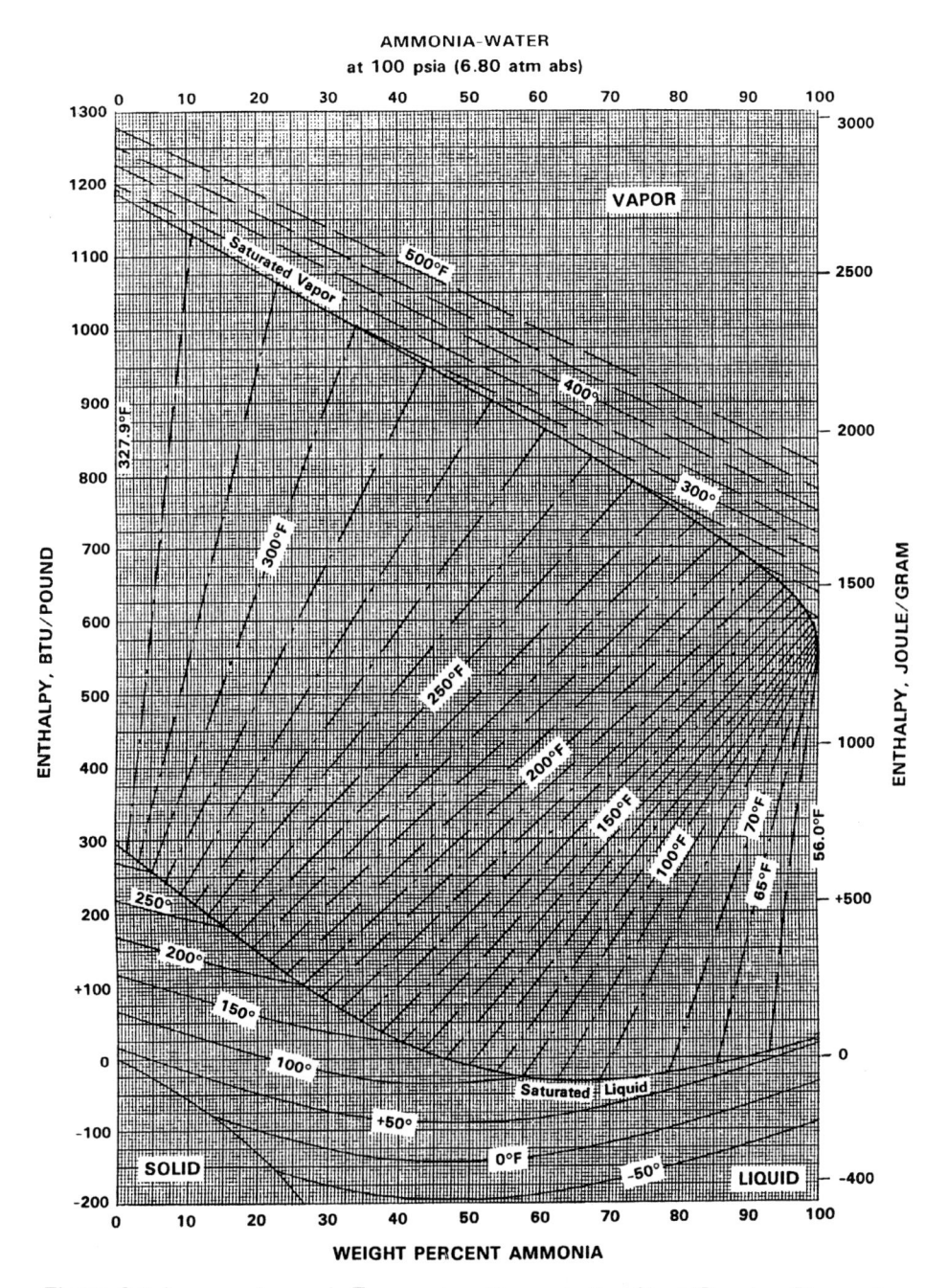

Figure A-2 Aqueous Ammonia Temperature-Concentration Chart (Ponchon Diagram)
From *Ammonia Absorption Refrigeration in Industrial Processes*, by Marcel Bogart. Table E-5.
Copyright ©1981 by Gulf Publishing Company, Houston, TX. Used with permission.
All rights reserved.

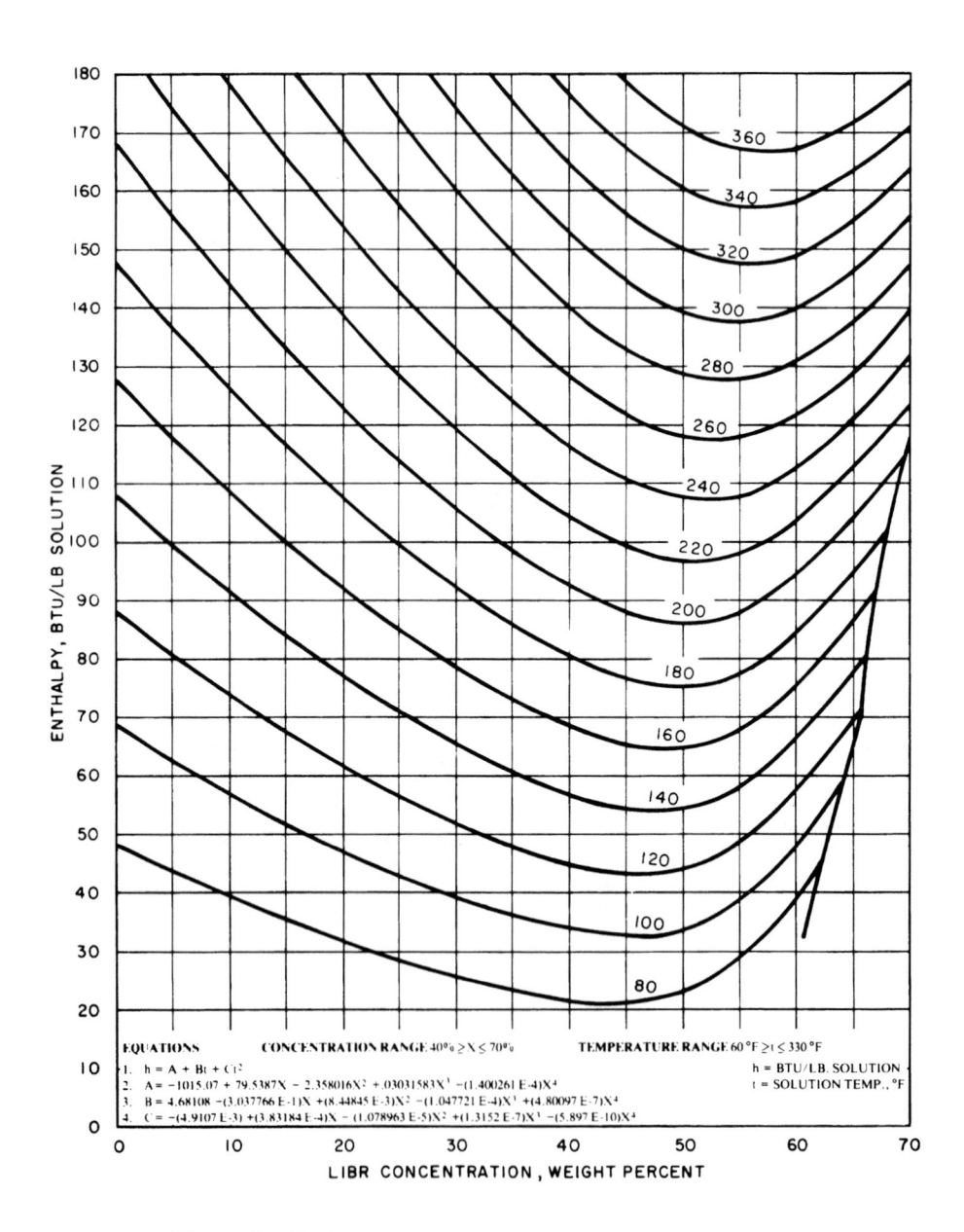

Figure A-3 Equilibrium Chart for Aqueous Lithium Bromide Solutions
Source: ASHRAE. 1993. Handbook of Fundamentals. American Society of Heating,
Refrigerating and Air-Conditioning Engineers, Inc., Atlanta, GA, Chapter 17, Figure 39.

TABLE A-4
Lithium Bromide Property Table

Refrigerant Temperature ($t' = °F$) and Enthalpy ($h = Btu/lb$) of Lithium Bromide Solutions

Temp., ($t' = °F$)		0	10	20	30	40	45	50	55	60	65	70
80	t'	80.0	78.2	75.6	70.5	60.9	53.5	42.1	28.6	13.8	−0.2#	−11.6#
	h	48.0	39.2	31.8	25.6	21.6	21.2	23.0	28.7	38.9	52.7#	37.1#
100	t'	100.0	98.1	95.3	89.9	79.6	71.8	60.0	46.1	30.9	16.2#	3.8#
	h	68.0	56.6	47.0	38.7	33.2	32.1	33.2	38.2	47.8	61.1	75.1
120	t'	120.0	117.9	114.9	109.2	98.3	90.1	77.9	63.6	48.1	32.7	19.1#
	h	87.9	73.6	61.7	51.7	44.7	43.0	43.6	48.0	56.9	69.4	83.0
140	t'	140.0	137.8	134.6	128.5	117.1	108.5	95.8	81.2	65.2	49.1	34.4#
	h	107.9	91.0	77.0	65.1	56.5	54.1	54.1	57.9	66.1	78.0	91.1#
160	t'	160.0	157.7	154.3	147.9	135.8	126.8	113.8	98.7	82.3	65.6	49.7#
	h	127.9	108.2	92.0	78.2	68.1	65.1	64.7	67.9	75.4	86.6	99.2#
180	t'	180.0	177.5	173.9	167.2	154.5	145.1	131.7	116.2	99.5	82.0	65.1#
	h	147.9	125.4	107.9	91.9	80.4	76.6	75.3	77.7	84.6	95.1	107.2#
200	t'	200.0	197.4	193.6	186.5	173.3	163.5	149.6	133.7	116.6	98.5	80.4#
	h	168.0	143.4	123.3	105.3	92.1	87.4	85.9	87.8	94.1	104.0	115.6#
220	t'	220.0	217.2	213.3	205.8	192.0	181.8	167.5	151.3	133.7	114.9	95.7
	h	188.1	160.7	138.2	119.0	104.1	99.0	96.5	97.8	103.3	112.5	123.6
240	t'	240.0*	237.1*	232.9	225.2	210.7	200.2	185.4	168.8	150.9	131.4	111.0
	h	208.3*	178.4*	154.0	132.6	116.0	110.3	107.1	107.7	112.5	121.1	131.6
260	t'	260.0*	256.9*	252.6*	244.58	229.4	218.5	203.3	186.3	168.0	147.9	126.4
	h	228.6*	195.7*	169.1*	146.2*	128.1	121.6	117.6	117.6	121.6	129.5	139.5
280	t'	280.0*	276.8*	272.3*	263.8*	248.2*	236.8*	221.2	203.9	185.1	164.3	141.7
	h	249.1*	213.8*	185.1*	159.7*	140.0*	132.8*	128.1	127.5	130.6	137.9	147.6
300	t'	300.0*	296.7*	291.9*	283.1*	266.9*	255.2*	239.2*	221.4	202.3	180.8	157.0
	h	269.6*	231.6*	200.7*	173.5*	152.1*	144.1*	138.9*	137.3	139.8	146.5	155.5
320	t'	320.0*	316.5*	311.6*	302.5*	285.6*	273.5*	257.1*	238.9*	219.4	197.2	172.4
	h	290.3*	249.7*	216.3*	187.2*	164.2*	155.3*	149.5*	147.1*	148.8	154.9	163.4
340	t'	340.0*	336.4*	331.3*	321.8*	304.4*	291.9*	275.0*	256.4*	236.5*	213.7	187.7
	h	311.1*	267.9*	232.1*	201.0*	176.1*	166.6*	160.1*	157.0*	158.0*	163.5	171.0
360	t'	360.0*	356.2*	350.9*	341.1*	323.1*	310.2*	292.9*	274.0*	253.7*	230.1	203.0
	h	332.2*	286.1*	248.0*	214.9*	188.2*	178.0*	170.6*	166.8*	167.0*	171.9	178.3

* Extensions of data above 235°F are well above the original data and should be used with care.
# Supersaturated solution.

Source: ASHRAE. 1993. *Handbook of Fundamentals.* American Society of Heating, Refrigerating and Air-Conditioning Engineers, Inc., Atlanta, GA, Chapter 17, p. 17.81.

TABLE A-5
Aqueaous Ammonia Property Table

Specific Volume of Saturated Ammonia Solutions, ft^3/lb

Concentration, Ammonia (Mass basis)

Temp., °F	0	10	20	30	40	50	60	70	80	90	100	Temp., °F
20	0.0160	0.0165	0.0170	0.0176	0.0182	0.0190	0.0197	0.0207	0.0217	0.0230	0.0245	20
40	0.0160	0.0165	0.0171	0.0177	0.0184	0.0191	0.0200	0.0209	0.0221	0.0236	0.0253	40
60	0.0160	0.0166	0.0172	0.0178	0.0186	0.0193	0.0202	0.0212	0.0225	0.0241	0.0260	60
80	0.0161	0.0167	0.0173	0.0180	0.0188	0.0196	0.0205	0.0216	0.0230	0.0247	0.0267	80
100	0.0161	0.0168	0.0174	0.0182	0.0190	0.0198	0.0208	0.0220	0.0235	0.0254	0.0275	100
120	0.0162	0.0169	0.0176	0.0184	0.0192	0.0201	0.0211	0.0224	0.0241	0.0261	0.0284	120
140	0.0163	0.0170	0.0177	0.0185	0.0194	0.0203	0.0215	0.0229	0.0247	0.0268	0.0294	140
160	0.0164	0.0172	0.0179	0.0187	0.0196	0.0206	0.0219	0.0235	0.0254	0.0277	0.0306	160
180	0.0165	0.0173	0.0181	0.0190	0.0199	0.0210	0.0223	0.0241	0.0262	0.0286	0.0320	180
200	0.0166	0.0175	0.0183	0.0192	0.0202	0.0213	0.0228	0.0247	0.0270	0.0298	0.0338	200
220	0.0168	0.0176	0.0185	0.0194	0.0205	0.0217	0.0234	0.0255	0.0279	0.0312	0.0361	220

APPENDIX B
Algorithms for Charts and Computer Program

Summary Information on Graphs (includes general description from chapter 9 and equations used in regression).

All charts are based on a compilation of manufacturers' data from various sources. The data were graphed by means of a regression analysis using a curve-fit program. The equations obtained by this program are given. Limits are also given for an acceptable range of values to be used in these equations.

The curves in this appendix are in IP units only. The computer program has conversion constants built into it to calculate in IP or SI units.

Figure 9-2

This figure (Figure 9-2) contains information on steam pressure and temperature and is based on a LiBr absorption machine with constant output and at an inlet steam pressure of 12 psig (184 kPa) steam for single-effect machines and 115 psig (894 kPa) steam for double-effect machines. Chilled water at 44°F (7°C) is the output of the machine.

The equations relating steam temperature, °F (y-axis), to chilled-water temperature, °F (x-axis), are

Double-Effect
$$y = 720.5 - 12.90 \cdot (x) + 0.09953 \cdot (x)^2$$

Single-Effect
$$y = 406.2 - 5.486 \cdot (x) + 0.03583 \cdot (x)^2$$

Limits: (40°F < x < 60°F)

x = chilled water temperature, °F
y = steam temperature, °F.

Also, an equation that is not shown in a graph but is useful is the equation relating steam pressure, psig (y-axis), to chilled-water temperature, °F (x-axis).

Double-Effect
$y = 801.1248 - 24.5778 \cdot x + 0.2035 \cdot x^2$

Single-Effect
$y = 91.5137 - 2.8606 \cdot x + 0.0219 \cdot x^2$

Limits: (40°F < x < 60°F).
x = chilled-water temperature, °F
y = steam pressure, psig.

Figure 9-3

This figure (Figure 9-3) is a graphical representation of the characteristics of the LiBr absorption machines currently available. Given a "heat input" (y-axis) calculated with Equation 9-1, "cooling capacity" (x-axis) can be determined for a generic absorption machine at a desired condition. The limits of this figure—the upper and lower boundaries—are the high and low steam pressure limits for the economic operation of a typical absorption system. These limits and the equations to these curves are

Single-Effect Machine:
a = 12 psig (184 kPa)
$x = (y - 0.233963)/17{,}573.064$ Limits: (150 tons < x < 2,500 tons)

b = 4 psig (129 kPa)
$x = (y - 0.145169)/16{,}298.508$ Limits: (150 tons < x < 2,200 tons)

Double-Effect Machine:
c = 144 psig (1,099 kPa)
$x = (y - 0.680634)/11{,}727.408$ Limits: (150 tons < x < 1,750 tons)

d = 65 psig (550 kPa)
$x = (y - 0.157432)/9{,}728.976$ Limits: (150 tons < x < 1,300 tons)

x = cooling capacity, tons
y = heat input, Btuh.

Figures 9-4 and 9-5

These figures (Figures 9-4 and 9-5) are used to determine the steam pressure correction factor, K_6, needed for Equation 9-4. Given a percentage of steam pressure (x-axis) and the entering condenser water temperature, the percentage of rated output capacity (y-axis)

APPENDIX B

can be determined. Note: 100% of rated steam pressure is 12 psig (184 kPa) for single-effect machines, and 115 psig (894 kPa) for double-effect machines. The equations to these curves are

Single-Effect Machine (Figure 9-4):
- A: $y = 70.1373 + 0.4122 \cdot x - 0.0011 \cdot x^2$
- B: $y = 69.6597 + 0.4043 \cdot x - 0.0010 \cdot x^2$
- C: $y = 66.0167 + 0.4607 \cdot x - 0.0012 \cdot x^2$
- D: $y = 62.3463 + 0.4846 \cdot x - 0.0011 \cdot x^2$
- E: $y = 59.2649 + 0.3641 \cdot x + 0.0003 \cdot x^2$

Limits: (0% < x < 117%)

Double-Effect Machine (Figure 9-5):
- A: $y = 74.2665 + 0.3864 \cdot x - 1183.4217/x$
- B: $y = 57.1307 + 0.5169 \cdot x - 917.4971/x$
- C: $y = 50.2667 + 0.6036 \cdot x - 1141.6352/x$
- D: $y = 51.2725 + 0.6520 \cdot x - 1683.715/x$
- E: $y = -3.4755 + 1.0819 \cdot x - 714.8106/x$

Limits: (38% < x < 112%)

x = percentage of steam pressure
y = percentage of rated output capacity.

Figures 9-6 and 9-7

These figures (Figures 9-6 and 9-7) are used to determine the leaving chilled-water correction factor, K_7, used in Equation 9-6. Given a leaving chilled-water temperature (x-axis) and the entering condenser water temperature, the load ratio (actual tons/nominal tons) (y-axis) can be determined. The equations to these curves are

Single-Effect Machine (Figure 9-6):
- 75°F $y = 1/(.0003 \cdot (x - 70.8864)^2 + 0.5949)$
- 80°F $y = 1/(.0003 \cdot (x - 73.5720)^2 + 0.6232)$
- 85°F $y = 3.0598 - 137.6866/x + 2074.3171/x^2$
- 90°F $y = 0.0201 \cdot x$
- 95°F $y = 1.1008 \cdot \exp[(x - 68.9956)^2/-1424.6368]$

Limits: (40°F < x < 60°F)

Double-Effect Machine (Figure 9-7):
- 75°F $y = 0.0278\,x$
- 80°F $y = -0.1777 + 0.029\,x$
- 85°F $y = -0.2178 + 0.0278\,x$

90°F $y = -0.3618 + 0.0275\, x$
95°F $y = -0.4473 + 0.0267\, x$

Limits: (40°F < x < 60°F)

x = leaving chilled-water temperature, °F
y = load ratio

Figures 9-9 and 9-10

These figures (Figures 9-9 and 9-10) represent the percent of design heat (y-axis) to the percent of design load (x-axis) for different entering condenser water temperatures. The equations to these curves are

Single-Effect Machine (Figure 9-9):

55°F $y = 9.7338 \cdot 2.7182818^{\frac{(\ln x - 1.5466)^2}{4.914}}$ Limits: (10 < x < 50)

65°F $y = 10.0028 \cdot 2.7182818^{\frac{(\ln x - 1.0526)^2}{5.9528}}$ Limits: (10% < x < 80%)

75°F $y = 13.384 \cdot 2.7182818^{\frac{(\ln x - 1.5007)^2}{4.9767}}$ Limits: (10% < x < 100%)

85°F $y = 14.5005 \cdot 2.7182818^{\frac{(\ln x - 1.4425)^2}{5.1559}}$ Limits: (10% < x < 100%)

95°F $y = 17.6766 \cdot 2.7182818^{\frac{(\ln x - 1.7158)^2}{4.991}}$ Limits: (10% < x < 80%)

Double-Effect Machine (Figure 9-10):
55°F $y = 1.99 \cdot 1.0074^x \cdot x^{0.6101}$ Limits: (0% < x < 60%)
65°F $y = 2.1751 \cdot 1.0068^x \cdot x^{0.637}$ Limits: (0% < x < 85%)
75°F $y = 2.9036 \cdot 1.0077^x \cdot x^{0.5784}$ Limits: (0% < x < 100%)
85°F $y = 3.549 \cdot 1.0072^x \cdot x^{0.5683}$ Limits: (0% < x < 100%)

x = percent of design load
y = percent of design heat input.

Figure 10-3

Figure 10-3 is in SI units. The reference documentation was prepared in SI units and therefore the following equations are in SI units. Figure 10-2 was developed by converting the desired IP parameters from SI and then reconverting the data to IP.

APPENDIX B

Figure 10-3 represents the parameters used in the sizing of aqueous ammonia absorption refrigeration systems. The equations used in Figure 10-3 are

$$COP_{max} = f_{0(Ts)} + f_{1(Ts)} (T_e) + f_{2(Ts)} (1/T_e) + f_{3(Ts)} [1/(T_e^2)]$$

$$T_{g,opt} = g_{0(Ts)} + g_{1(Ts)} (T_e) + g_{2(Ts)} (T_e)^2 + g_{3(Ts)} (T_e)^3$$

$$f_{j(Ts)} = a_{0j} + a_{1j} (T_s) + a_{2j} (T_s)^2 + a_{3j} (T_s)^3$$

$$g_{j(Ts)} = b_{0j} + b_{1j} (T_s) + b_{2j} (T_s)^2 + b_{3j} (T_s)^3$$

T_s = ammonia condensate temperature from the condenser
T_e = evaporator temperature
T_g = strong solution (weak aqua) temperature.

The coefficients for the previous equations used to develop Figure 10-3 are

	COP_{max}		$T_{g,opt}$
$a_{0,0}$	6.55132e−1	$b_{0,0}$	−7.26209
$a_{1,0}$	−1.5648e−2	$b_{1,0}$	2.91205
$a_{2,0}$	9.30096e−4	$b_{2,0}$	−1.86364e−2
$a_{3,0}$	−1.36564e−5	$b_{3,0}$	1.85319e−4
$a_{0,1}$	4.41648e−3	$b_{0,1}$	−2.58944
$a_{1,1}$	−2.77987e−4	$b_{1,1}$	1.24279e−1
$a_{2,1}$	1.98907e−5	$b_{2,1}$	−4.42468e−3
$a_{3,1}$	−2.66527e−7	$b_{3,1}$	4.29658e−5
$a_{0,2}$	−6.19823	$b_{0,2}$	−6.52162e−2
$a_{1,2}$	5.77539e−2	$b_{1,2}$	6.98244e−3
$a_{2,2}$	7.47419e−3	$b_{2,2}$	−2.38075e−4
$a_{3,2}$	−1.36508e−4	$b_{3,2}$	2.26658e−6
$a_{0,3}$	−3.27816e+1	$b_{0,3}$	−8.99504e−4
$a_{1,3}$	4.40948e−1	$b_{1,3}$	9.70080e−5
$a_{2,3}$	1.80843e−2	$b_{2,3}$	−3.27227e−6
$a_{3,3}$	−3.80018e−4	$b_{3,3}$	3.05454e−8

The above equations are good for the following ranges:

Ts = 5°C to 45°C   Te = −55°C to −5°C
   = 40°F to 115°F   = −65°F to 25°F.

APPENDIX C
Input Instructions for Computer Program

The ASHSORPT program provides a computerized method for calculating the characteristics of an absorption system using the procedure described in chapters 8 through 10. The program can be used to quickly evaluate many different absorption system options.

The program is started on an IBM-compatible computer by typing "ASHSORPT" at the DOS prompt. After the title screen is displayed, the input screen shown in Figure C-1 appears. When AAR is the selected machine type, only those values that apply to ammonia absorption refrigeration appear on the screen.

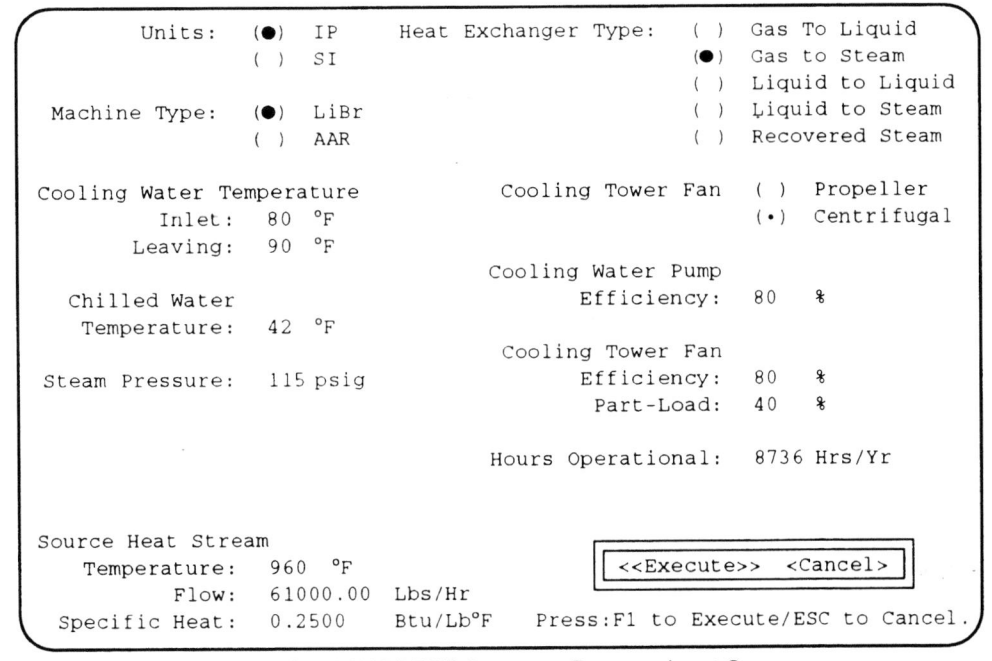

```
            Units:     (●)  IP      Heat Exchanger Type:  ( )  Gas To Liquid
                       ( )  SI                            (●)  Gas to Steam
                                                          ( )  Liquid to Liquid
     Machine Type:     (●)  LiBr                          ( )  Liquid to Steam
                       ( )  AAR                           ( )  Recovered Steam

   Cooling Water Temperature              Cooling Tower Fan  ( )  Propeller
             Inlet:   80  °F                                 (•)  Centrifugal
           Leaving:   90  °F
                                          Cooling Water Pump
        Chilled Water                             Efficiency:  80  %
         Temperature:  42  °F
                                          Cooling Tower Fan
      Steam Pressure:  115 psig                  Efficiency:  80  %
                                                  Part-Load:  40  %

                                          Hours Operational:  8736 Hrs/Yr


   Source Heat Stream
         Temperature:    960    °F             ┌─────────────────────────┐
                Flow:    61000.00  Lbs/Hr      │ <<Execute>>   <Cancel>  │
       Specific Heat:    0.2500   Btu/Lb°F     └─────────────────────────┘
                                         Press:F1 to Execute/ESC to Cancel.
```

Figure C-1 ASHSORPT Computer Program Input Screen

Arrow keys move the cursor around the screen. Where numerical input is not required, the space bar is used to select the desired option. Calculations can be made by selecting <Execute> or by hitting <F1>. Selecting <Cancel> or hitting <ESC> terminates the program.

The following inputs are used to estimate absorption machine parameters for *LiBr applications*.

Units: SI for metric units, IP for inch-pound units.

Machine Type: LiBr or AAR.

Heat Exchanger Type: Gas to liquid, gas to steam, liquid to liquid, liquid to steam, or direct use of recovered steam.

Cooling Water Temperatures

Inlet: Temperature of the water entering the absorption machine (not the temperature of the water entering the cooling tower). It must be lower than the temperature of the water leaving the absorption machine.

Leaving: Temperature of the water leaving the absorption machine. This is typically 10°F to 20°F (6°C to 12°C) higher than the water temperature entering the absorption machine.

Chilled-Water Temperature: Temperature of chilled water leaving absorption machine.

Steam Pressure: If a steam absorption machine is to be used, this is the pressure of the steam entering the generator. For hot water analysis, this is not used.

Hot Water Temperatures

Inlet: If a gas-to-liquid or liquid-to-liquid heat exchanger is selected, this is the temperature of the water entering the generator. For gas-to-steam or liquid-to-steam heat exchangers, this value is not used.

Outlet: If a gas-to-liquid or liquid-to-liquid heat exchanger is selected, this is the temperature of the water leaving the generator. Typically, this is 10°F to 60°F (6°C to 33°C). For gas-to-steam or liquid-to-steam heat exchangers, this value is not used.

Source Heat Stream

Temperature: Temperature of liquid or gas stream used as a source.

Flow: Flow of recovered fluid (lb/h [kg/s] for recovered gas, gpm [kg/s] for recovered liquid).

Specific Heat: Specific heat of source heat stream in Btu/lb•°F (J/kg•K).

Cooling Tower Fan: Centrifugal or propeller fan.

Cooling Water Pump

Efficiency: Efficiency of cooling water pump. Expressed as a whole number (50% expressed as 50, not 0.50).

Cooling Tower Fan

Efficiency: Efficiency of cooling tower fan. Expressed as a whole number (50% expressed as 50, not 0.50).

Part Load: Percent of time the cooling tower is in operation. Expressed as a whole number (50% expressed as 50, not 0.50).

APPENDIX C

Hours Operational: Number of hours the absorption machine will run in a single year.

The following inputs are used to estimate absorption machine parameters for *AAR applications*.

Units: SI for metric units, IP for inch-pound units.
Machine Type: LiBr or AAR.
Heat Exchanger Type: Gas to liquid, gas to steam, liquid to liquid, liquid to steam, or direct use of recovered steam.
Cooling Water Temperatures
   Inlet: Temperature of cooling water entering the absorption machine.
   Chilled-Water Temperature: Temperature of chilled water leaving absorption machine.
Source Heat Stream
   Temperature: Temperature of liquid or gas stream used as a source.
   Flow: Flow of recovered fluid (lb/h [kg/s] for recovered gas, gpm [kg/s] for recovered liquid).
   Specific Heat: Specific heat of source heat stream in Btu/lb·°F (J/kg·K).

Output for a LiBr calculation is shown in Figure C-2. AAR output can be seen in Figure C-3.

```
Equivalent Steam Pressure    115 psig     Machine Type:    Double-Effect Steam LiBr
Recovered Heat                            Exhaust Gas Leaving Temperature:   392  °F
Total:    8669113            Btuh
Flow:        8114            Lb/hr        Cooling Tower:    Centrifugal

Cooling Output: 795          tons

Steam Pressure Corr. Factor:  1.00        Corrected Output:   798 tons

Water Temp. Corr. Factor:     1.04        Corrected Output:   767 tons (Nominal)


Cooling Water Flow:    4.5 gpm/ton        Total Heat Rejection:    17885844 Btuh

Abs. Machine:  0/018  kW/ton       Energy Use:  120585    kWh      Demand:      14 kW
Cooling Wat.:  0.100  bhp/ton      Energy Use:  647312kWh Demand:               74 kW
Cooling Twr.:  0.230  bhp/ton      Energy Use:  595527    kWh      Demand:     170 kW

Total Energy Use: 1363425   kWh        Demand:   258 kW   Demand/Ton:  0.325 kW/ton

First Costs —Abs. Machine:$ 350683   Heat Rej.: $ 174884   Heat Rec.:  $ 202847

Total First Cost:$   728414          Cost per Unit Cooling:    $916.33/ton
```

Figure C-2 ASHSORPT Output — LiBr Case

```
COP Max:       0.59
Generator Temp:    196.6      °F
Heat Recovered:    10065000   Btuh

QOut:   394    tons
```

Figure C-3 ASHSORPT Output—AAR Case

Any calculation result that is out of range of the program will be displayed as a zero (0) on the output screen.

APPENDIX D
Existing Applications Listing

On the following pages are tables showing existing applications of absorption units utilizing recovered heat. This listing is a small sampling of absorption systems currently installed. The purpose of the listing is to provide scenarios for the application of absorption machines.

One-stage ammonia absorption refrigeration (AAR) systems are in Tables D-1 (IP units) and D-5 (SI units). Tables D-2 (IP units) and D-6 (SI units) contain information on two-stage AAR systems. Single-effect LiBr systems are listed in Tables D-3 (IP units) and D-7 (SI units). Tables D-4 (IP units) and D-8 (SI units) contain information on double-effect LiBr systems. The data shown are for the peak-load design conditions. See the *Remarks* column for site-specific information.

APPLICATION GUIDE FOR ABSORPTION COOLING/REFRIGERATION USING RECOVERED HEAT

Table D-1
One-Stage AAR Systems, IP Units

	Site Location	Duty	No. of Units	Manufacturer/ Model	Size (tons)	Type	Source	Steam Pressure (psig)	Inlet Temp. (°F)	Outlet Temp. (°F)	Flowrate (lb/hr)
1	Agip-Casirate, Italy	refrigeration	2	Borsig	500	oil			482	365	204,600
2	Boskovich Farms	refrigeration	1	Lewis Energy Systems — 860	860	steam	cogeneration		394	390	33,398
3	California Site Using Ammonia	tube ice machines	1	Borsig	800	steam	cogeneration	intermediate	360	250	32,000
4	Chevron U.S.A.	refrigeration	1	Lewis Energy Systems — 250	250	steam	wtr treatment system		292	270	324,870
5	Lenzing AG, Austria	refrigeration	1	Borsig	1,700	steam	turbine exhaust	superheated	356	288	39,700
6	Lenzing AG, Austria	process	1	Borsig	1,700	steam	turbine exhaust	superheated	356	288	39,700
7	Pictsweet	refrigeration	1	Lewis Energy Systems — 1000	890	steam	cogeneration		366	358	4m350

Recovered Heat (column group above)

		Chilled Fluid			Cooling Water				Costs ($)	
	Type	Inlet Temp. (°F)	Outlet Temp. (°F)	Flow (lb/hr)	Inlet Temp. (°F)	Outlet Temp. (°F)	Flow (gpm)	Installation	Remarks
1	gas	15	−40	15,915	82	102	2,070	3,000,000	Installation costs include everything
2	ammonia	−10	−24	varies	75	95	3,790		Frame 5000 gas turbine
3	ammonia	100	−6	3,365	80	95	2,900		Vogt MSG boiler surface heat recovery
4	ammonia	41	92		82	102	1,005		Waste heat from water treatment system
5	NaOH brine	41	27	700,000	81	97	7,045	8,000,000	Crystallization of process fluid
6	NaOH brine	41	27	700,000	71	97	7,45	8,000,000	Brine cooling, costs include everything
7	ammonia	−19	−33	17,970	75	97	4,500		Food processing facility

Note: Bottom half of table is continuation of top half.

162

Appendix D

Table D-2
Two-Stage AAR Systems, IP Units

Site Location	Duty	No. of Units	Manufacturer/ Model	Size (tons)	Type	Source
1 Bechtel Inc.	dewaxing process	1	Lewis Energy Systems 675/515	675 515	steam steam	turbine exhaust turbine exhaust
2 Sunlaw	cold storage	2	Borsig	1,000	steam	cogeneration

Recovered Heat

Steam Pressure (psig)	Inlet Temp. (°F)	Outlet Temp. (°F)	Flowrate (lb/hr)
	308	310	21,600
	287	285	17,281
105			15,000

Chilled Fluid

Type	Inlet Temp. (°F)	Outlet Temp. (°F)	Flow (lb/hr)
1 ammonia	-13	-15	14,000
ammonia	10	0	13,000
2 ammonia	100	92	3,365

Cooling Water

Inlet Temp. (°F)	Outlet Temp. (°F)	Flow (gpm)
87	105	3,000
87	105	2,600
82	102	1,005

Costs ($)

Installation	Remarks
	First stage
	Second stage
	Running 7 yrs, very little maintenance

Note: Bottom half of table is continuation of top half.

163

Table D-3
Single-Effect LiBr Systems, IP Units

	Site Location	Duty	No. of Units	Manufacturer/ Model	Size (tons)	Type	Source	Steam Pressure (psig)	Inlet Temp. (°F)	Outlet Temp. (°F)	Flowrate (lb/hr)
1	Big Six Towers	hvac	1	Carrier 16JB018	174	steam	cogeneration	12			3,132
2	Dow, Freeport, TX	process/ refrigeration	2	Trane ABSC-16C	1,660	steam	process	15	250	160	15,000
3	Fowler High School	hvac	1	Carrier 16JB018	80	hot water	cogeneration		215	180	79 gpm
4	Frehafer Hall, West Lafayette, Indiana	hvac	1	Trane ABSC14C4	1,200	steam	backpress, turbine	15	350	180	22,000
5	Henninger High School	hvac	1	Carrier 16JB014	81	hot water	cogeneration		220	176	58 gpm
6	Hoechst Celanese Chemical, Bishop Plant	process	1	Trane ABSC-12A1	1,200	steam		15			
7	Hoechst Celanese Chemical, Pasadena Plant	process/hvac	4	Trane C16C	1,600	steam	process	12	270	245	
8	Hoechst Celanese Chemical, Pasadena Plant	process process	1 1	Trane C16C Carrier 16JB	1,200 450	steam steam	process process	12 12	270	245	
9	Hoechst Celanese Chemical, Pasadena Plant	process process	1 1	Trane C12C Trane C08	1,200 800	steam steam	process process	12 12	270	245	
10	Lincoln School	hvac	1	Carrier 16JB014	81	hot water	cogeneration		220	176	58 gpm
11	Little Co. of Mary Hospital	hvac	1	Carrier 16JB054	572	steam	cogeneration	12			10,470
12	Maine South High School	hvac	1	Carrier 16JB057	600	hot water	cogeneration		260	229	645 gpm
13	Occidental Chemical	process	2	Trane ABSC-12C	1,200	hot water	condensate		150		

APPENDIX D

14	Oneida Madison Boces	hvac	1	Carrier 16JB024	115	hot water	cogeneration		220	185	113 gpm
15	Oscar Mayer Foods Corp.	hvac	1	Carrier 16JB068-604		steam		60			
16	Resurrection Hospital	hvac	1	Carrier 16JB028-604-15	501	steam	cogeneration	60 12			9,018
17	St. Vincent's Hospital	hvac	1	Carrier 16JB068	745	steam	cogeneration	14			13,633
18	University of Illinois	hvac	67	Carrier (most)	500*	steam	cogeneration	12			
19	University of Missouri	hvac	50	Mixed Carrier, Trane, York	700**	steam	extracted, 60 psi	15			
20	University of Northern Colorado	hvac	2	Arida HTHW→ Stm.→CHW	30	hot water	cogeneration	270	230		
		hvac	5	Carrier HTHW/ 16JB	280***	hot water	cogeneration	270	230	220 gpm
		hvac	3	Trane 1A series, 2B series	350***	hot water	cogeneration	270	230	220 gpm
21	University of Wisconsin-Madison	hvac	1	Trane A5B-3	565	steam	generator	12			10,600
22	Valparaiso University—Christ College	hvac	1	Carrier 16J014-60	60	hot water	cogeneration		260		
	Gellerson Eng. Bldg.	hvac	1	Trane ABSC	160	hot water	cogeneration		260		

*   machine sizes range from 100 tons to 750 tons
**   machine sizes range from 100 tons to 1,000 tons
*** machine sizes range from 240 tons to 320 tons (Carrier)
        machine sizes range from 300 tons to 408 tons (Trane)

Note: Table D-3 continues on next page.

Table D-3
Single-Effect LiBr Systems, IP Units *(continued)*

	Type	Chilled Fluid Inlet Temp. (°F)	Chilled Fluid Outlet Temp. (°F)	Flow (gpm)	Cooling Water Inlet Temp. (°F)	Cooling Water Outlet Temp. (°F)	Flow (gpm)	Costs ($) Installation	Costs ($) Operating	Maintenance	Remarks
1	water	54	44	418	85	102	522				Per unit
2	water	54	41	3,060	75	86	9,960	600,000	3,000,000	60,000	2 5-hp pumps
3	water	54	44	192	85	91	839				Cost per machine, 1982 dollars
4	water	54	40	2,042	80	86	4,000	350,000			
5	water	58	53	400	85	91	720				
6	water	45	40	5,000	90	105	5,800				Run all @ 65%, 3 handle load for
7	water	52	42	2,400	93	102	5,300		160,000	18,000	maintenance
8	water	52	42	2,200	93	102	5,200		40,000	4,500	
	water	52									
9	water		42	2,200	93	102	5,200		40,000	4,500	2 5-hp pumps
	water		42							4,500	Installation includes equipment
10	water	58	53	400	85	91	720				cost only
11	water	54	44	1,374	85	101	2,060	120,000			
12	water	63	48	950	85	102	2,000	800,000	6,000	3,500	Maintenance w/Trane, 2 20-hp motors.
13	water	67	55	2,400	87	105					4.5/kWh
14	water	50	45	467	85	93	864				2 5-hp pumps
15		51	42						7,000	13,000	Cannot claim as waste heat
	water	51	42						chemicals		4 BAC 250-ton cooling towers
16	water	55	45	1,202	85	101	1,470				
17	water	57	45	1,500	85	102	2,500				14 hp motors total, 480 V
18	water		45–52		85	95			$.067/ ton-hr	$.003/ ton-hr	Operating includes elec. & steam

APPENDIX D

19	water	55	45		85	95		700,000	Derated due to dirty cooling water, 1 30-hp pump
20	water (20% glycol)	55 55	45 45	700 700	80 80	95 95	1,000 1,000	1,750,000 1,050,000	Heat recovery steam generators from privately owned system, Flows are approximate.
21	water	55	45	1,356	85	104	1,900		11 hp motors total
22	water water	55	44 44					2,000	Heat recovery equipment, Baird from Louisiana, Gas turbine exhaust

167

Table D-4
Double-Effect LiBr Systems, IP Units

	Site Location	Duty	No. of Units	Manufacturer/ Model	Size (tons)	Type	Source	Steam Pressure (psig)	Inlet Temp. (°F)	Outlet Temp. (°F)	Flowrate (lb/hr)
1	AT&T Bell Labs	hvac	1	Carrier 16JT150	1,500	steam	cogeneration	125			15,000
2	Brooklyn Hospital #1	hvac	1	Carrier 16JT090	900	steam	cogeneration	114			8,910
3	Brooklyn Hospital #2	hvac	1	Carrier 16JT841	350	steam	cogeneration	105			3,465
4	Freehafer Hall, West Lafayett, Indiana	hvac	2	Trane ABTD10C	900	steam	backpress, turbine	125	550		11,665
5	Rice University	hvac	2	Trane ABTC-10C	1,000	steam	cogeneration	123	400		12,300
		hvac	1	Hitachi HAU-W-22G	1,000	steam	cogeneration	123	400		15,000
6	Providence Memorial Hospital	hvac	1	Carrier 16JT857	500	steam	cogeneration	114			4,850

APPENDIX D

	Chilled Fluid					Cooling Water					Costs ($)			
Type	Inlet Temp. (°F)	Outlet Temp. (°F)	Flow (gpm)		Inlet Temp. (°F)	Outlet Temp. (°F)	Flow (gpm)		Installation	Operating	Maintenance	Remarks
1 water	54	42	3,000		85	95	6,623					8 kW in pump energy
2 water	57	45	1,800		85	100	2,700					
3 water	57	45	700		85	100	1,050					
4 water	54	40			80	86			350,000			Cost per machine, 1982 dollars
5 water	54	44	2,400		85	100	3,300		250,000		30,000	2 gas turbines, heat recovery, ERI
water	52	44	4,500		85	95	6,700		250,000		30,000	25,000 lb/hr @ 240 psig, ERI 350 psig @ 550F, 25,000 lb/hr
6 water	54	44	1,263		85	96	2,000		25,000			Absorption machine cost only

Note: Bottom half of table is continuation of top half.

169

Table D-5
One-Stage AAR Systems, SI Units

Recovered Heat

Site Location	No. of Units	Duty	Manufacturer/ Model	Size (kW$_T$)	Type	Source	Steam Pressure (kPa)	Inlet Temp. (°C)	Outlet Temp. (°C)	Flowrate (kg/hr)
1 Agip-Casirate, Italy	2	refrigeration	Deufsche Babcock-Borsig AG	1,750	steam	cogeneration		250	185	92,800
2 Boskovich Farms	1	refrigeration	AAAR-860	3,025	steam	cogeneration		201	199	15,150
3 California Site Using Ammonia	1	tube ice machines	Borsig, Berlin '89	2,800	steam	cogeneration	inter- mediate	180	120	14,515
4 Chevron U.S.A.	1	refrigeration	AAAR-250	900	steam	wtr. treatment system		145	130	14,760
5 Lenzing AG, Austria	1	refrigeration	Deufsche Babcock-Borsig AG	6,000	steam	turbine exhaust	super- heated	180	140	18,000
6 Lenzing AG, Austria	1	process	Deufsche Babcock-Borsig AG	6,000	steam	turbine exhaust	super- heated	180	140	18,000
7 Pictsweet	1	refrigeration	AAR-1000	3,100	steam	cogeneration		185	180	1,975

Chilled Fluid / Cooling Water / Costs

	Type	Inlet Temp. (°C)	Outlet Temp. (°C)	Flow (kg/hr)	Inlet Temp. (°c)	Outlet Temp. (°c)	Flow (L/s)	Instal- lation	Oper- ating	Maint- enance	Remarks
1	gas	−9	−31	7,220	28	39	130	3,000,000			Installation costs include everything
2	ammonia	−23	−21	varies	24	35	240				Frame 5000 gas turbine
3	ammonia	38	33	1,530	27	35	5				Vogt MSG boiler surface heat recovery
4	ammonia	5	−3	317,520	28	39	65				Waste heat from water treatment system
5	NaOH brine	5	−3	317,520	27	36	445	8,000,000			Crystallization of process fluid
6	NaOH brine	5	−3	317,520	22	36	445	8,000,000			Brine cooking, costs include everythig
7	ammonia	−28	−36	8,150	24	36	285				Food processing facility

Note: Bottom half of table is continuation of top half.

APPENDIX D

Table D-6
Two-Stage AAR Systems, SI Units

Site Location	Duty	No. of Units	Manufacturer/ Model	Size (kW$_T$)	Type	Source	Steam Pressure (kPa)	Inlet Temp. (°C)	Outlet Temp. (°C)	Flowrate (kg/hr)
1 Bechtel Inc.	dewaxing process	1	AAAR 675-615	2,375 2,160	steam steam	turbine exhaust turbine exhaust		153 142	154 141	9,800 7,840
2 Sunlaw	cool storage	2	Borsig	3,500	steam	cogeneration	825			6,800

Recovered Heat

	Chilled Fluid			Cooling Water			Costs ($)			
Type	Inlet Temp. (°C)	Outlet Temp. (°C)	Flow (kg/hr)	Inlet Temp. (°c)	Outlet Temp. (°c)	Flow (kg/hr)	Instal- lation	Oper- ating	Maint- enance	Remarks
1 ammonia ammonia	−25 −12	−26 −18	6,350 5,900	30	40	1,360				First stage Second stage
2 ammonia		−40								Running 7 yrs, very little maintenance

Note: Bottom half of table is continuation of top half.

171

Table D-7
Single-Effect LiBr Systems, SI Units

	Site Location	Duty	No. of Units	Manufacturer/ Model	Size (kW$_T$)	Type	Source	Steam Pressure (kPa)	Inlet Temp. (°C)	Outlet Temp. (°C)	Flowrate (kg/hr)
1	Big Six Towers	hvac	1	Carrier 16JB018	610	steam	cogeneration	184			1,420
2	Dow, Freeport, TX	process/ refrigeration	2	Trane ABSC-16C	5,840	steam	process	205	120	70	6,800
3	Fowler High School	hvac	1	Carrier 16JB018	280	hot water	cogeneration		102	82	5 L/s
4	Frehafer Hall, West Lafayette, Indiana	hvac	1	Trane ABSC14C4	4,220	steam	backpress. turbine	205	177	82	9,980
5	Henninger High School	hvac	1	Carrier 16JB014	285	hot water	cogeneration		104	80	4 L/s
6	Hoechst Celanese Chemical, Bishop Plant	process	1	Trane ABSC-12A	4,220	steam		205			
7	Hoechst Celanese Chemical, Pasadena Plant	process/hvac	4	Trane C16C	5,630	steam	process	184	132	118	
8	Hoechst Celanese Chemical, Pasadena Plant	process process	1 1	Trane C16C Carrier 16JB	4,220 1,580	steam steam	process process	184 184	132	118	
9	Hoechst Celanese Chemical, Pasadena Plant	process process	1 1	Trane C12C Trane C08	4,220 2,845	steam steam	process process	184 184	132	118	
10	Lincoln School	hvac	1	Carrier 16JB014	285	hot water	cogeneration		104	80	4 L/s
11	Little Co. of Mary Hospital	hvac	1	Carrier 16JB054	2,010	steam	cogeneration	184			4,750
12	Maine South High School	hvac	1	Carrier 16JB057	2,110	hot water	cogeneration		127	109	41 L/s
13	Occidental Chemical	process	2	Trane ABSC-12C	4,220	hot water	condensate		66		

Recovered Heat

APPENDIX D

14	Oneida Madison Boces	hvac	1	Carrier 16JB024	400	hot water	cogeneration		104	85	7 L/s
15	Oscar Mayer Foods Corp.	hvac	1	Carrier 16JB068-604		steam	cogeneration	515		
			1	Carrier 16JB028-604-15		steam		515		
16	Resurrection Hospital	hvac	1	Carrier 16JB054	1,760	steam	cogeneration	184		4,090
17	St. Vincent's Hospital	hvac	1	Carrier 16JB068	2,620	steam	cogeneration	198		6,185
18	University of Illinois	hvac	67	Carrier (most)	1,700*	steam	cogeneration	184		
19	University of	hvac	50	Mixed Carrier, Trane, York	2,500**	steam	extracted, 60 psi	205		
20	University of Northern Colorado	hvac.	2	Arida HTHW→Stm.→CHW	105	hot water	cogeneration		132	110	
		hvac	5	Carrier HTHW/1,000*** 16JB		hot water	cogeneration		132	110	14 L/s
		hvac	3	Trane 1A series,1,200*** 2B series		hot water	cogeneration		132	110	14 L/s
21	University of Wisconsin-Madison	hvac	1	Trane A5B-3	2,000	steam	generator	184		4,800
22	Valparaiso University—Christ College	hvac	1	Carrier 16J014-60	210	hot water	cogeneration		127		
	Gellerson Eng. Bldg.	hvac	1	Trane ABSC	560	hot water	cogeneration	127		

* machine sizes range from 350 kWT to 2600 kWT
** machine sizes range from 350 kWT to 3500 kWT
*** machine sizes range from 840 kWT to 1125 kWT (Carrier)
   machine sizes range from 1050 kWT to 1435 kWT (Trane)

Note: Table D-7 continues on next page.

173

Table D-7
Double-Effect LiBr Systems, SI Units *(continued)*

Type	Chilled Fluid Inlet Temp. (°C)	Outlet Temp. (°C)	Flow (L/s)	Cooling Water Inlet Temp. (°C)	Outlet Temp. (°C)	Flow (L/s)	Costs ($) Installation	Operating	Maintenance	Remarks
1 water	12	7	25	29	39	35				Per unit
2 water	12	5	545	24	30	1,630	600,000	3,000,000	60,000	2 5-hp pumps
3 water	12	7	12	29	33	55				Cost per machine, 1982 dollars
4 water	12	4	130	27	30	250	350,000			
5 water	14	12	25	29	33	45				
6 water	7	4	315	32	41	365				
7 water	11	6	150	34	39	335		160,000	18,000	Run all @ 65%, 3 handle load for maintenance
8 water	11	6	140	34	39	330		40,000	4,500	
9 water		6	140	34	39	330		40,000	4,500	
water		6							4,500	
10 water	14	12	25	29	33	35	120,000			2 5-hp pumps
11 water	12	7	90	29	39	130				Installation includes equipment cost only
12 water	17	9	60	29	39	125	800,000	6,000	3,500	Maintenance w/Trane. 2 20-hp motors, 4.5/kWh
13 water	19	13	150	31	41					
14 water	10	7	30	29	34	55				2 5-hp pumps
15 water	11	6						7,000 chemicals	13,000	Cannot claim as waste heat 4 BAC 250-ton cooling towers
water	11	6								
16 water	13	7	75	29	38	95				14 hp motors total, 480 V
17 water	14	7	95	29	39	160		$.067/ ton-hr	$.003/ ton-hr	Operating includes elec. & steam
18 water		7		29	35					

Appendix D

19	water	13	7		29	35		Derated due to dirty cooling water, 1 30-hp pump
20	water (20% glycol)	13	7	45	27	35	65	700,000	Heat recovery steam generators from privately owned system, Flows are approximate.
		13	7	45	27	35	65	1,750,000	
21	water	13	7	85	27	35		1,050,000	11 hp motors total
22	water	13	7		29	40	120	2,000	Heat recovery equipment, Baird from Louisiana, Gas turbine exhaust
	water		7						

APPLICATION GUIDE FOR ABSORPTION COOLING/REFRIGERATION USING RECOVERED HEAT

Table D-8
Double-Effect LiBr Systems, SI Units

Recovered Heat

	Site Location	Duty	No. of Units	Manufacturer/ Model	Size (kW$_T$)	Type	Source	Steam Pressure (kPa)	Inlet Temp. (°C)	Outlet Temp. (°C)	Flowrate (kg/hr)
1	AT&T Bell Labs	hvac	1	Carrier 16JT150	1,500	steam	cogeneration	960			6,805
2	Brooklyn Hospital #1	hvac	1	Carrier 16JT090	900	steam	cogeneration	890			4,040
3	Brooklyn Hospital #2	hvac	1	Carrier 16JT841	350	steam	cogeneration	825			1,570
4	Freehafer Hall, West Lafayett, Indiana	hvac	2	Trane ABTD10C	900	steam	backpress, turbine	965	290		5,290
5	Rice University	hvac hvac	2 1	Trane ABTC-10C Hitachi HAU-W-22G	1,000 1,000	steam steam	cogeneration cogeneration	950 950	205 205		5,580 6,805
6	Providence Memorial Hospital	hvac	1	Carrier 16JT857	500	steam	cogeneration	890			2,200

Chilled Fluid | Cooling Water | Costs ($)

	Type	Inlet Temp. (°C)	Outlet Temp. (°C)	Flow (L/s)	Inlet Temp. (°C)	Outlet Temp. (°C)	Flow (L/s)	Installation	Operating	Maintenance	Remarks
1	water	12	6	190	29	35	420				8 kW in pump energy
2	water	14	7	115	29	38	170				
3	water	14	7	45	29	38	65				
4	water	12	4	150	27	30	210	350,000			Cost per machine, 1982 dollars
5	water	12	7	150	29	38	210	250,000		30,000	2 gas turbines, heat recovery, ERI 25,000 lb/hr @ 240 psig, ERI 350
	water	11	7	285	29	35	430	250,000		30,000	psig @ 550F, 25,000 lb/hr
6	water	12	7	80	29	36	125	25,000			Absorption machine cost only

Note: Bottom half of table is continuation of top half.

176

GLOSSARY

Below are terms used throughout this guide. The definitions are from the *ASHRAE Handbook* series, the ASHRAE *Terminology of Heating, Ventilation, Air Conditioning, & Refrigeration* (1991), or as noted.

AAR: Ammonia absorption refrigeration (Bogart 1981, p. 30).

Absorbent: Material that, due to an affinity, extracts one or more substances from a liquid or gaseous medium with which it is in contact and that changes physically, chemically, or both, during the process. Calcium chloride is an example of a solid absorbent, while solutions of lithium chloride, lithium bromide, water, and ethylene glycol are examples of liquid absorbents (*Terminology*, p. 1).

Absorber: Absorbers are internally water-cooled tube bundles over which strong absorbent is sprayed or dripped in the presence of refrigerant vapor. The vapor is absorbed by the solution flowing over the outside of the tubes, releasing heat. The heat is removed by the cooling water (*1988 ASHRAE Equipment Handbook*, p. 13.3).

Absorption: The process whereby a porous material extracts one or more substances from an atmosphere, a mixture of gases, or a mixture of liquids (*Terminology*, p. 1).

Aqueous Ammonia System: The vaporizing refrigerant is ammonia and the absorbent liquid is water. The absorber produces a concentrated solution of ammonia in water, referred to as weak absorbent. The weak absorbent solution is separated by distillation into two streams—a liquid ammonia overhead product (recycled as refrigerant) and the water-rich bottoms (strong absorbent) used as the absorbent (Bogart 1981, pp. 30-31).

Coefficient of Performance (COP): A ratio of energy units of cooling or heating achieved to the same energy units required to power a refrigeration cycle used for cooling or heating. The heating COP may be designated as HCOP. Only the cooling COP is used in this guide.

Cogeneration: Sequential production of either electrical or mechanical power and useful thermal energy (heating or cooling) from a single energy form. See also *Electric Power Cogeneration* (*Terminology*, p. 16).

Concentrated Solution: See *Strong Absorbent*.

Condenser: Heat exchanger in which vapor is liquified by the rejection of heat to a heat sink (*Terminology*, p. 22).

In a LiBr machine: Internally water-cooled tube bundles located in the refrigerant vapor space over or near the generator (*1988 ASHRAE Equipment Handbook*, p. 13.3).

Droplet eliminators shield the condenser from solution carryover from the generator. In double-effect machines, the high-stage condenser is incorporated in the tube side of the low-stage generator.

Crystallization: The LiBr-water solution freezes. This reduces or stops the flow of solution in the absorption process (Reimann 1994).

Dilute Solution: See *Weak Absorbent*.

Ebullient Cooling: An engine-cooling process in which the coolant (usually water) undergoes a phase change within the engine. The water/steam mixture is less dense than water. The lower density mixture rises through the engine into a steam separator. The steam is available for use, while the water is naturally recirculated to the engine (Manufacturer p. 2–4).

Effect: An absorption machine with a nominal COP of 0.5 is a single-effect machine; with a nominal COP of 1.0, it is a double-effect machine, and so on.

Electric Power Cogeneration: Any of several processes that either use waste heat from generation of electricity to satisfy thermal needs or process waste heat in the steam generation of electricity. See also *Cogeneration* (*Terminology*, p. 38).

Evaporator: Part of a refrigerating system in which the refrigerant is evaporated, absorbing heat from the contacting heat source (*Terminology*, p. 40).

LiBr: Tube bundles over which liquid refrigerant is sprayed or dripped and evaporated. The liquid to be cooled passes inside the tubes (*1988 ASHRAE Equipment Handbook*, p. 13.3).

AAR: AAR systems utilize a flooded evaporator design, with ammonia from the AAR system flooding the shell side of a shell-and-tube heat exchanger. On the tube side, gas or liquid from the load is condensed or cooled (Shepherd 1994).

Generator: Section of an absorption machine where the refrigerant is separated from the solution using heat.

LiBr: Heat exchangers, usually of the tube bundle type, which are submerged in solution or arranged to accommodate a falling film of solution and heated by steam, hot liquids, or hot gases. The heat evaporates the refrigerant from solution (*1988 ASHRAE Equipment Handbook*, p. 13.3).

AAR: Vertical tanks finned on the outside to extract heat from the combustion products. Internally, a system of analyzer plates creates intimate counterflow contact between the vapor generated, which rises, and the absorbent, which descends (*1988 ASHRAE Equipment Handbook, p. 13.10*).

GLOSSARY

Heat Amplifier: Standard- or forward-cycle heat pump. Takes waste heat at low temperature and prime energy at high temperature and discharges medium temperatures. Also called a type I absorption heat pump or forward absorption heat pump (RCG/Hagler, Bailly, Inc. 1990).

Heat Recovery Boiler: A heat recovery device using exhaust heat (exhaust gas, hot discharge line, etc.) to generate appropriate heat energy to power an absorption system.

Indirect Fired Absorption Machine: An absorption machine that uses heat recovered from another process or heat cycle machine.

Lithium Bromide (LiBr)-Water System: An absorption system in which the water is the refrigerant and lithium bromide is the absorbent.

P-H Diagram: A chart used to graphically express refrigeration processes. Information in a P-H diagram includes pressure, temperature, enthalpy, entropy, and state.

PTX Chart: A chart to graphically express the LiBr absorption process. Information includes refrigerant vapor *pressure*, refrigerant *temperature*, solution *temperature*, and solution *concentration*.

Purge System: Devices that remove noncondensable gases. These gases, even when present in small quantities, raise the total pressure in the absorber, which causes an appreciable decrease in machine output (*1988 ASHRAE Equipment Handbook*, p. 13.3).

Rectifier (AAR): The rectifier comprises a spiral coil through which weak absorbent from the solution pump passes on its way to the absorber and generator. Vapor issuing from the generator is still partially laden with water vapor, which can be reduced to a negligible amount by cooling, using the rectifier.

  Some type of packing is included to assist counterflow contact between condensate from the coil (which is refluxed to the generator) and the vapor (which continues on to the condenser) (*1988 ASHRAE Equipment Handbook*, p. 13.11).

Refrigerant: Fluid used for heat transfer in a refrigerating system. The fluid absorbs heat at low temperature and low pressure and transfers heat at a higher temperature and higher pressure, usually involving changes of the state of the fluid (*Terminology*, p. 77).

Solution: A mixture of two elements. In absorption systems, the common solutions are a LiBr-water solution and a water-ammonia solution.

Strong Absorbent: Solution with a high affinity for refrigerant (*1988 ASHRAE Equipment Handbook*, p. 13.1).
  LiBr: referred to as a concentrated solution.
  AAR: referred to as a weak aqua.

Strong Aqua: See *Weak Absorbent*.

Temperature Amplifier: Reverse-cycle heat pump. Takes waste heat at medium temperature and discharges heat at both a higher and lower temperature. Also called a type II absorption heat pump, heat transformer, or temperature booster (RCG/Hagler, Bailly, Inc. 1990).

Weak Absorbent: Solution with a low affinity for refrigerant (*1988 ASHRAE Equipment Handbook*, p. 13.1).
    LiBr: referred to as a dilute solution.
    AAR: referred to as a strong aqua.

Weak Aqua: See *Strong Absorbent*.

REFERENCES

ASHRAE. 1988. *1988 ASHRAE handbook—Equipment*. Atlanta: American Society of Heating, Refrigerating and Air-Conditioning Engineers, Inc.

ASHRAE. 1991. *ASHRAE terminology of heating, ventilation, air conditioning, and refrigeration*. Atlanta: American Society of Heating, Refrigerating and Air-Conditioning Engineers, Inc.

Bogart, M. 1981. *Ammonia absorption refrigeration in industrial processes*. Houston: Gulf Publishing Co.

Manufacturer. 1988. *Waukesha cogeneration handbook*. Waukesha Dresser.

RCG/Hagler, Bailly, Inc. 1990. *Opportunities for industrial chemical heat pumps in process industries*. Prepared for the U.S. Department of Energy, contract no. DE-AC01-87CE40762, subcontract no. 3090-1.

Reimann, B. 1994. Personal communication (phone).

Shepherd, J. J. 1994. Personal communication (letter).

BIBLIOGRAPHY

AGCC. 1992. *Natural gas cooling equipment guide.* Arlington, VA: American Gas Cooling Center.

Akridge, J. M., and J. H. Sitz. 1989. Technical tour of the Georgia Power corporate headquarters building. *ASHRAE Journal,* December, pp. 34–41.

Andberg, J. W., and G. C. Vliet. 1983. Design guidelines for water-lithium bromide absorbers. *ASHRAE Transactions* 89(2): 220–231.

Anon. 1985. The quest for smaller units. *ASHRAE Journal,* July, pp. 18–23.

ARI. 1992a. *Standard 560, Absorption water chilling and water heating packages.* Arlington, VA: Air-Conditioning and Refrigeration Institute.

ARI. 1992b. *Standard 590, Positive displacement compressor water-chilling packages.* Arlington, VA: Air-Conditioning and Refrigeration Institute.

ARI. 1992c. *Standard 550, Centrifugal and rotary screw water-chilling packages.* Arlington, VA: Air-Conditioning and Refrigeration Institute.

ARI. 1993. Annual sales figures. Arlington, VA: Air-Conditioning and Refrigeration Institute.

ASHRAE. 1988. *1988 ASHRAE handbook—Equipment.* Atlanta: American Society of Heating, Refrigerating and Air-Conditioning Engineers, Inc.

ASHRAE. 1991. *1991 ASHRAE handbook—HVAC applications.* Atlanta: American Society of Heating, Refrigerating and Air-Conditioning Engineers, Inc.

ASHRAE. 1992. *1992 ASHRAE handbook—HVAC systems and equipment.* Atlanta: American Society of Heating, Refrigerating and Air-Conditioning Engineers, Inc.

ASHRAE. 1993. *1993 ASHRAE handbook—Fundamentals.* Atlanta: American Society of Heating, Refrigerating and Air-Conditioning Engineers, Inc.

ASHRAE. 1994. *1994 ASHRAE handbook—Refrigeration.* Atlanta: American Society of Heating, Refrigerating and Air-Conditioning Engineers, Inc.

ASME. 1994. *Proceedings of the International Absorption Heat Pump Conference,* AES-Vol. 31. New Orleans, LA. New York: American Society of Mechanical Engineers, Advanced Energy Systems Division.

Ball, H. D. 1980. The consumption of energy and of capital in dehumidification systems. *ASHRAE Transactions* 86(1): 1031–1036.

Beaird Industries, Inc. 1990. *Installation, operation and maintenance instructions—Maxim TRP heat recovery silencers.* Bulletin No. 143 90TRP. Shreveport, LA: Beaird Industries, Inc.

Beaird Industries, Inc. 1992. *Heat recovery application manual.* Bulletin No. 1430592. Shreveport, LA: Beaird Industries, Inc.

Best, R., W. Rivera, I. Pilatowsky, and F. A. Holland. 1990. Thermodynamic design data for absorption heat transformers—Part four: Operating on ammonia-lithium nitrate. *Heat Recovery Systems & CHP* 10(5/6): 539–548.

Bogart, M. 1981. *Ammonia absorption refrigeration in industrial processes.* Houston: Gulf Publishing Co.

Boyen, J. L. 1980. *Thermal energy recovery*, 2d ed. New York: John Wiley & Sons.

Briggs, S. W. 1971. Concurrent, crosscurrent, and countercurrent absorption in ammonia-water absorption refrigeration. *ASHRAE Transactions* 77(1): 171–175.

Buffington, R. M. 1949. Qualitative requirements for absorbent-refrigerant combinations. *Refrigerating Engineering* 4(April): 343.

Cacciola, G., G. Restuccia, and N. Giordano. 1990. Economic comparison between absorption and compression heat pumps. *Heat Recovery Systems & CHP* 10(5/6): 499–507.

Cain Industries. 1994. Personal correspondence with Jim Rozanski.

Carrier. 1975. *Hermetic absorption liquid chillers.* Catalog No. 16JB-3P. Syracuse, NY: Carrier Corporation.

Carrier. 1993. *Double-effect hermetic absorption liquid chiller.* Catalog No. 16JT-1PD. Syracuse, NY: Carrier Corporation.

Caterpillar. 1984. *On-site power generation handbook.* No. LEBX4457. Milwaukee, WI: Caterpillar, Inc.

Caterpillar. 1992a. *G3400 gas engine, engine performance.* Catalog No. LEBQ2024. Milwaukee, WI: Caterpillar, Inc.

Caterpillar. 1992b. *G3500 gas engine, engine performance.* Catalog No. LEBQ2023. Milwaukee, WI: Caterpillar, Inc.

Caterpillar. 1993. *Generator sets, application and installation guide.* No. LEBX3377. Milwaukee, WI: Caterpillar, Inc.

Ciambelli, P., and V. Tufano. 1990. Coupling a single-stage absorption heat transformer with finite heat sources/sinks. *Heat Recovery Systems & CHP* 10(5–6): 549–553.

Clements, J. R., J. R. Macanliss, and P. R. Steinway. 1984. Packaged gas-fueled cogeneration system for hospitals annual report (September 1983–September 1984). Martin Tractor Company for Gas Research Institute.

Coellner, J. A. 1980. Energy conservation techniques for use in rotary solid sorption dehumidification systems. *ASHRAE Transactions* 86(1): 1022–1026.

Collier, R. K., Jr., D. Novosel, and W. M. Worek. 1990. Performance analysis of open-cycle desiccant cooling systems. *ASHRAE Transactions* 96(1): 1262–1268.

Coltec Industries. 1993. *Cogeneration.* Fairbanks Morse Engine Division, File No. 3019E 2.5M-4/93.

Columbia Gas Distribution Companies. 1993. *Double-effect absorption gas heat pump—Residential and commercial prospectus.* Columbus, OH: Columbia Gas Distribution Companies.

Davidson, W. F., and D. C. Erickson. 1988. Absorption heat pumping for district heating now practical. *ASHRAE Transactions* 94(1): 707–715.

BIBLIOGRAPHY

Dorgan, C. B. 1994. Personal communication with Trane, York, and multiple HVAC contractors nationwide (phone).

Dorgan, C. E., and J. S. Elleson. 1993. *Design guide for cool thermal storage*. Atlanta: American Society of Heating, Refrigerating and Air-Conditioning Engineers, Inc.

Dorgan Associates, Inc. 1994. Survey of HVAC contractors across the United States.

DRC. 1994. Tri-services automated cost engineering system (TRACES) cost estimating analysis, system design. Niceville, FL: Delta Research Corp.

Eisa, M.A.R., P. J. Diggory, and F. A. Holland. 1987. Experimental studies to determine the effect of differences in absorber and condenser temperatures on the performance of a water-lithium bromide absorption cooler. *Energy Conversion Management* 27(2): 253–259.

Fallek, M. 1985. Parallel flow chiller-heater. *ASHRAE Transactions* 91(2B): 2095–2102.

Fallek, M. 1986. Absorption chillers for cogeneration applications. *ASHRAE Transactions* 92(1B): 141–147.

Furlong, J. 1994. Personal communication (letter).

Gilbert and Associates. 1993. *CFCs and electric chillers, selection of large capacity water chillers in the 1990s (revision 1)*. EPRI TR-100537, R1. Palo Alto, CA: Gilbert and Associates.

GRI. 1987. *How gas cools (or, apples can "fall" up)*. PB88-123096. Chicago: Gas Research Institute.

GRI. 1990. Triple-effect absorption chiller. Tech profile 1290LB12000. Chicago: Gas Research Institute.

Grimm, N. R., and R. C. Rosaler, eds. 1990. *Handbook of HVAC design*. Chapter 41, "Absorption Chillers," by N. J. Cassimatis. New York: McGraw-Hill.

Grossman, G. 1991. Absorption heat transformers for process heat generation from solar ponds. *ASHRAE Transactions* 97(1): 420–427.

Grossman, G., and H. Perez-Blanco. 1982. Conceptual design and performance analysis of absorption heat pumps for waste heat utilization. *ASHRAE Transactions* 88(1): 451–466.

Guinn, G. R. 1991. Planning cogeneration systems. *ASHRAE Journal*, February, pp. 18–22.

Harkins, H. L. 1987. Maximum power cogeneration for the pulp and paper industry. *IEEE Conference Record of the Annual Pulp and Paper Industry Technical Conference*, pp. 147–151.

Harris, K. J. 1987. Consider absorption cooling to tap cogeneration waste heat. *Power*, August, pp. 47–48.

Herold, K. E., and R. Radermacher. 1989. Absorption heat pumps. *Mechanical Engineering*, August, pp. 68–73.

Holldorff, G. M., and W. F. Malewski. 1987. Operational experience in cogeneration plants with refrigeration supply to low-temperature cold storage. *ASME, International Gas Turbine Institute* 1: 287–295.

Hufford, P. E. 1990. Cogeneration gold—Recovered heat. *ASME, Petroleum Division, PD* 28: 47–51.

Hufford, P. E. 1991. Absorption chillers maximize cogeneration value. *ASHRAE Transactions* 97(1): 428–433.

Hufford, P. E. 1992. Absorption chillers improve cogeneration energy efficiency. *ASHRAE Journal* 34(3): 46–53.

Ikeuchi, M., T. Yumikura, E. Ozaki, and G. Yamanaka. 1985. Design and performance of a high-temperature-boost absorption heat pump. *ASHRAE Transactions* 91(2B): 2081–2094.

Inoue, N., H. Iizuka, Y. Nimomiya, K.-I. Wantabe, and T. Acki. 1994. *COP evaluation for advanced ammonia-based absorption cycles.* AES No. 31, pp. 1–6.

Jacob, X., L. F. Albright, and W. H. Tucker. 1969. Factors affecting the coefficient of performance for absorption air-conditioning systems. *ASHRAE Transactions* 75(1): 103–109.

Jain, P. C., and G. K. Gable. 1971. Equilibrium property data equations for aqua-ammonia mixtures. *ASHRAE Transactions* 77(1): 149–151.

Jennings, B. H., and F. P. Shannon. 1938. The thermodynamics of absorption refrigeration. *The Journal of the ASRE*, May, pp. 333–336.

Jiexiu, D., T. Minduan, and Y. Jingchang. 1992. Computer prediction and experimental study of operation behavior for absorption heat transformer. *Thermodynamics and the Design, Analysis, and Improvement of Energy Systems.* AES-Vol. 27/HTD-Vol. 228, pp. 149–152. New York: American Society of Mechanical Engineering.

Jung, S.-H., C. Sgamboti, and H. Perez-Blanco. 1993. *An experimental study of the effect of some additives on falling film absorption.* AES-Vol. 31, International Absorption Heat Pump Conference, pp. 49–55. New York: American Society of Mechanical Engineers.

Kuhlenschmidt, D., and R. H. Merrick. 1983. An ammonia-water absorption heat pump cycle. *ASHRAE Transactions* 89(1B): 215–219.

Lenard, J. 1994. Thermal energy storage increases Cogen abilities. *ASME News*, February, p. 5.

LESI. 1994. Product catalogs. North Salt Lake City, UT: Lewis Energy Systems, Inc.

Linnell, C. J. 1985. Cost-effective small-scale CHP. *Energy Digest*, pp. 13–18. (Based on paper given to Importer '85 Conference by Linnell.)

Machielsen, C.H.M. 1990. Research activities on absorption systems for heating, cooling and industrial use. *ASHRAE Transactions* 96(1): 1577–1581.

Means, R. S. 1993. *Means mechanical cost data, 1994, 17th annual edition.* Kingston, MA: R. S. Means.

Moyer, C. B., J. A. Nicholson, and P. T. Overly. 1986. Opportunity evaluation for gas engine and gas turbine cogeneration/absorption refrigeration systems in industrial applications. Prepared for Gas Research Institute by the Acurex Corporation.

Murray, J. G. 1993. Purge needs in absorption chillers. *ASHRAE Journal* 35(10): 40–47. Also published in *ASHRAE Transactions* 99(1): 1485–1494.

Oh, M. D., S. C. Kim, Y. L. Kim, and Y. I. Kim. 1994. Cycle analysis of air-cooled, double-effect absorption heat pump with parallel flow type. AES-Vol. 31, International Absorption Heat Pump Conference, pp. 117–123. New York: American Society of Mechanical Engineers.

Ouimette, M. S., and K. E. Herold. 1994. *Performance modeling of triple-effect absorption chiller.* AES #31, pp. 233–241.

Pande, A. C. 1983. Total energy center Wellington Hospital design of systems and services. *ASHRAE Transactions* 89(2A): 153–163.

Patnaik, V., H. Perez-Blanco, and W. A. Ryan. 1993. A simple analytical model for the design of vertical tube absorbers. *ASHRAE Transactions* 99(2): 69–80.

Pawelski, M. J. 1994. Personal communication (meeting).

Perez-Blanco, H., and R. Radermacher. 1986. Absorption: An update. *ASHRAE Journal*, November, pp. 25–26.

Plzak, W. 1994. Personal communications (phone).

Purtell, R. F. 1989. Cogeneration, chillers and cool storage. *ASHRAE Journal* 31(2): 26–29.

Radermacher, R. 1984. Heat pump cycles with nonazeotropic refrigerant mixtures in thermodynamic diagrams. *ASHRAE Transactions* 90(2A): 166–174.

Radermacher, R., S. A. Klein, and D. A. Didion. 1983. Investigation of the part-load performance of an absorption chiller. *ASHRAE Transactions* 89(1A): 205–213.

Rane, M. V., and D. C. Erickson. 1994. Advanced absorption cycle: Vapor exchange GAX. AES #31, pp. 25–32.

RCG/Hagler, Bailly, Inc. 1990. Opportunities for industrial chemical heat pumps in process industries. Prepared for the U.S. Department of Energy, contract no. DE-AC01-87CE40762, subcontract no. 3090-1.

Reimann, B. 1994. Personal communication (phone and letter).

Rockenfeller, U. 1994. Personal communication (meeting and phone).

Rohrer, W. M., Jr., and K. G. Kreider. 1979. *Industrial and institutional waste heat recovery.* Chapter 1, "Sources and Uses of Waste Heat." Noyes Data Corp.

Ryan, W. 1994. Personal communication (letter).

Sakamoto, K., M. Nagao, and T. Katayama. 1990. Cogeneration systems for telecommunication power plants. *INTELEC—International Telecommunications Energy Conference (Proceedings)*: 438–442. New York: Institute of Electrical and Electronics Engineers.

Sami, S. M. 1990. Hybrid model for evaluating the performance of open-cycle absorption systems for waste heat recovery. *Heat Recovery Systems & CHP* 10(5/6): 573–582.

Scalabrin, G., and G. Scaltriti. 1985. A new energy saving process for air dehumidification: Analysis and applications. *ASHRAE Transactions* 91(1A): 426–441.

Shepherd, J. J. 1994. Personal communication (letter, fax, and phone).

Smith, W. P., Jr. 1990. *Absorption refrigeration: Performance at the USA sites.* Presented at Annual RE Technology Exchange-BASF, June.

Smith, W. P., Jr. 1994. Personal communication (letter and phone).

Smith, W. P., Jr., R.C. Erickson, W.A. Liegois, and C.E. Dorgan. 1992. *Cogeneration technology.* Course text, College of Engineering, EPD Department. Madison: University of Wisconsin.

SnyderGeneral Corporation. 1991. *Double effect absorption water chillers 100 to 1500 tons.* Catalog 975. Minneapolis, MN: SnyderGeneral Corporation, McQuay Division.

STI. 1983. *Cogeneration systems.* T75J/887/20M. San Diego, CA: Solar Turbines, Inc.

Stoecker, W. F., and L. D. Reed. 1971. Effect of operating temperatures on the coefficient of performance of aqua-ammonia refrigerating systems. *ASHRAE Transactions* 77(1): 163–169.

Tamblyn, R. T. 1985. College Park thermal storage experience. *ASHRAE Transactions* 91(1B): 947–955.

Trane. 1972. *Absorption operation maintenance, applications engineering manual.* No. AM-FND4-772. La Crosse, WI: The Trane Company.

Trane. 1981. *Two stage absorption cold generator 385 to 1000 tons.* Catalog No. ABS-D-2, April. La Crosse, WI: The Trane Company.

Trane. 1985. *Absorption refrigeration.* Air Conditioning Clinic No. 2803-11-677. La Crosse, WI: The Trane Company.

Trane. 1989. *Single stage absorption cold generator.* Catalog No. ABS-DS-1, March. La Crosse, WI: The Trane Company.

U.S.T. 1993. *Kawasaki gas turbine, technical data manual.* Cincinnati, OH: U.S. Turbines.

Vicatos, G., and J. Gryzagoridis. 1994. Unpublished research, Department of Mechanical Engineering. Cape Town, South Africa: University of Cape Town.

Vliet, G. C., and F. B. Cosenza. 1991. Absorption phenomena in water-lithium bromide films. Japanese Absorption Heat Pump Conference, Tokyo, Japan. *Conference Proceedings.*

Walker, D. H., I. P. Krepchin, E. C. Poulin, R. L. Demler, and S. J. Hynek. 1985. Gas-fueled cogeneration for supermarkets: Phase I—Final report. Prepared for Gas Research Institute. Contract No. 5083-243-0931.

Walker, D. H., S. J. Hynek, B. N. Barck, D. L. Fischbach, R. D. Tetreault, and N. B. Longo. 1986. A new approach to cogeneration in the supermarket industry. *Proceedings of the Intersociety Energy Conversion Engineering Conference 21st, ACS,* pp. 154–158.

Waukesha Dresser. 1988. *Waukesha cogeneration handbook.*

Waukesha Dresser. 1993. Product catalogs.

Whitlow, E. P., and J. S. Swearingen. 1958. An improved absorption refrigeration cycle. *Gas Age* 122(9): 19–22.

Wood, B. D., N. S. Berman, K. Kim, and D.S.C. Chau. 1993. Heat transfer additives for absorption cooling system fluids. Prepared for Gas Research Institute, #93/0149.

Yong, L.-F., and R. M. Nelson. 1990. An expert system to select heat exchangers for waste heat applications. *ASHRAE Transactions* 96(1): 1539–1548.

York. 1993. *Two stage steam absorption chillers.* Catalog No. 155.19-EG2P (493). York, PA: York Corporation.

Yumikura, T., M. Ikeuchi, E. Ozaki, G. Yamanaka, and T. Arai. 1989. Experimental studies on characteristics of a two-stage absorption heat transformer. *ASHRAE Transactions* 95(2): 175–183.